S0-BKM-776

IDEAS IN
SOIL AND PLANT NUTRITION

by
Joe Traynor

Front cover photo: Grower Tim Indart surveys barley-wild oat field, Clovis, Ca. (yellowing is due to frost).

Back cover photo: Friant canal water, Madera, Ca. irrigation district.

Photos by Hartt Porteous, Fresno, Ca.

Library of Congress Catalog Card No. 80-82531
ISBN 0-9604704-0-9

Published by Kovak Books
P.O. Box 1422
Bakersfield, CA 93302

CONTENTS

INTRODUCTION

This book attempts to condense some of the vast amount of information on soil and plant nutrition in a meaningful way. It is also intended to bring out points that aren't stressed in other sources. The perspective of the book is from the point of view of the man in the field. The points stressed are points that have been important to me in my own field experience.

The word "fieldman" is used often herein, as in "the fieldman should be aware of". "Fieldman" could mean the representative of a chemical company but could also mean a farmer or consultant. Look at the word fieldman as being "the man in the field", whether he (she) be chemical company representative, farmer, consultant, farm advisor, student or professor.

For ease of reading, no notations are made in the text. The references and notes in the "Reference Notes" section at the end of the book document or discuss points made in the text. The reference notes are not necessary to the comprehension of the text, but are there for anyone interested in pursuing a particular point. That many of the references are drawn from trade publications (rather than from scientific journals) is not a condescension but a compliment to the material and quality of writing to be found in such publications.

For those wishing more detailed information on a specific topic, there are general references listed at the end of each chapter. A "General Reference" section at the end of the book lists twelve references on soil and plant nutrition; the fieldman that is seriously interested in the subject should either own or have access to most of these references.

Joe Traynor

Bakersfield, California
June 1980

PART I:

GENERAL CONSIDERATIONS

KOVAK BOOKS
P. O. Box 1422
Bakersfield, California 93302

NEW BOOK:

Ideas in
SOIL AND PLANT NUTRITION

BY
JOE TRAYNOR

This 120 page book summarizes much of the available information on soil and plant nutrition in a meaningful way. Written by a professional consultant with over 20 years of field experience.

- PRICE SCHEDULE -

Quantity	Price per copy
1 - 9	$5.00 each
10 - 19	$4.50
20 or more	$4.00

Use the coupon below, or send order to Kovak Books, P.O. Box 1422, Bakersfield, Ca. 93302.

- -

To: Kovak Books
P.O. Box 1422
Bakersfield, Ca. 93302

Enclosed is $ _____ (check or money order) for _____

copies of **Ideas in Soil and Plant Nutrition** @ $ _____ per copy.

Name_____

Address_____

_____

1. IT'S THE WATER!

A working knowledge of water quality is indispensable to the practicing agriculturist-fieldman. Water quality influences initial crop selection and then crop management throughout the life of the crop. In diagnosing field problems, water quality is one of the first things that should be evaluated. The professional agriculturist should be able to quickly scan a water analysis report and pick out pertinent information.

A major obstacle to quick interpretation of water analysis results is that data can be reported in one of several different units. The fieldman must be able to easily convert from one unit to another for interpretation. All that is required is the memorizing (or keeping in a handy place) a few basic rules and conversion factors as outlined in the following.

Reading Data

Water analysis data are reported in one of three units: ppm, mg/l or meq/l (parts per million, milligrams per liter or milliequivalents per liter). ppm and mg/l are identical terms for interpretive purposes. To change from ppm to meq/l (or vice versa), the equivalent weight of the chemical constituent in question must be known, then:

$$\text{equivalent weight X meq/l} = \text{ppm (or mg/l)}$$
$$\text{OR}$$
$$\text{ppm (or mg/l)} \div \text{equivalent weight} = \text{meq/l}$$

EQUIVALENT WEIGHTS

Cations (+ charged)		Anions (− charged)	
Ca (calcium)	20.0	Cl (chloride)	35.5
Mg (magnesium)	12.2	NO_3 (nitrate)	62
Na (sodium)	23.0	SO_4 (sulfate)	48
Ca+Mg	19.2	CO_3 (carbonate)	30
(assume 90% is Ca)		HCO_3 (bicarbonate)	61

p 14 gyp

3

Examples:

1. How many ppm is 3.3 meq/l Ca?
   3.3 X 20.0 = 66.0 ppm

2. How many meq/l is 50 ppm Na?
   $\dfrac{50}{23}$ = 2.17 meq/l

In addition, nitrate can be reported either as nitrate (NO_3) or nitrate-nitrogen (NO_3-N), and sulfate can be reported as SO_4 or SO_4-S. To convert nitrate to nitrate-N, divide by 4.4 (50 ppm NO_3 = 11.4 ppm NO_3-N). To convert sulfate to sulfate-sulfur, divide by 3.0 (48 ppm SO_4 = 16 ppm SO_4-S).

Another handy fact in water analysis interpretation is that an acre foot of water weighs 2,716,435 lbs. (or 2.7 million lbs.). Thus water with a nitrate-nitrogen content of 10 ppm (10 lbs. per million lbs.) contains 27 lbs. of N per acre foot. With a normal crop usage of 3 to 4 acre feet of water per season, such a water supplies a considerable portion of a crop's fertilizer needs.

Electrical Conductivity (EC)

EC or Electrical Conductivity is usually reported in millimhos per centimeter (mmhos/cm) and measures the total salt content of water. Multiplying EC by 640 gives a good approximation of the total ppm salt or total dissolved solids (TDS). Multiplying EC by 10 gives a good approximation of the total cation or anion concentration in meq/l.

The symbol "K" is sometimes used in place of EC and EC is sometimes written as EC X 10^{-3} which is the same as mmhos/cm. EC X 10^{-6} is micromhos/cm and can be converted to mmhos by multiplying by 1000. EC X 10^{-5} can be converted to mmhos by multiplying by 100.

Water Classification by Toxic Constituents

The agricultural suitability of water is determined by the concentration of three constituents: sodium, chloride and boron.

A tentative classification is given below:

Table 1. - **Classification of Water by Toxic Constituents**[*]

Classification	- - Concentration of constituents ---		
	meq/l Na	meq/l Cl	ppm B
Excellent	less than 4	less than 3.5	less than 0.3
Good to very good	4 - 6	3.5 - 4.5	0.3 - 0.5
Fair to good	6 - 7	4.5 - 5.5	0.5 - 1.0
Poor to fair	7 - 9	5.5 - 7.5	1.0 - 2.0
Very poor and possibly injurious	9 - 12	7.5 - 10.0	2.0 - 3.0
Very poor and usually injurious	12 - 16	10 - 14	3.0 - 4.0
Injurious to suitable if used over a prolonged period of time	over 16	over 14	over 4.0

The above classification does not consider total salt content, however in virtually all agricultural situations, if a water is excessive in total salt it will be excessive in sodium and/or chloride. (A low total salt content should be taken into consideration because permeability problems can occur with low salt water. Permeability problems are probable when EC is below 0.2 mmhos/cm and possible when EC is between 0.2 and 0.5. Ammendment application will offset permeability problems caused by low salt water).

Many researchers refrain from classifying waters because both water management and environmental conditions influence water suitability. The four main conditions under which a very poor quality water could be satisfactorily used are:

1. very low evapo-transpiration (cool, overcast climate)
2. drip irrigation
3. a well drained soil kept continually well supplied with water
4. cultivation of only very tolerant crops

The ameliorating effect of environmental conditions on water quality was brought home to me on a visit to a broccoli field in Baja California. Prior to the visit I had seen the water data which showed both sodium and chloride in the 10 to 12 meq/1 range. I was confident of seeing considerable leaf burning and plant stunting and was quite surprised when I saw an excellent field of broccoli about to be harvested. The climate was cool (next to the

[*] the salt and boron tolerance of many crops is given in USDA Handbook 60

5

ocean with many foggy days), the soil well drained and kept well supplied with water (furrow irrigation) proving that the effective classification of a water can be changed by both climate and management.

Other Toxic Constituents

Fieldmen in the southwestern U.S. should be aware that some wells contain toxic concentrations of lithium. Also, sewage waters in some areas contain toxic amounts of fluoride, cadmium and other heavy metals.

The classification in Table 1 does not consider a fourth constituent - bicarbonate. (some waters also contain carbonate, which is similar to bicarbonate in it's adverse effects; any reference herein to bicarbonate is meant as carbonate + bicarbonate). It is known that bicarbonate in water has direct adverse effects on some plant species (see following chapter). As more research on bicarbonate is published and as more bicarbonate sensitive plants become known, bicarbonate concentration will become an important criterion for determining water suitability for certain crops. Fortunately, the amount of bicarbonate in water can be reduced to safe levels by the addition of acid. Although waters are not currently classified by bicarbonate content as they are by sodium, chloride and boron content, bicarbonate concentration is considered in the calculation of a fourth classification criterion discussed below.

Water Classification - A Fourth Criterion

The 3 toxic constituents (sodium, chloride and boron) upon which water classification is based are essentially unchangeable - i.e., there is no practical way of reducing high levels of these constituents to safe levels; (under greenhouse culture, water with an excess of sodium, chloride or boron can be run through exchange resins or reverse osmosis systems but this is not yet practical for large scale agriculture). A fourth criterion estimates the sodium hazard of waters and considers the ratio of calcium + magnesium (Ca+Mg) to sodium (Na) and carbonate + bicarbonate ($CO_3 + HCO_3$).

A low amount of Ca+Mg in relation to Na and $CO_3 + HCO_3$ can

6

cause soil sealing and consequent poor water infiltration due to a buildup of sodium in the soil. Calcium and magnesium reduce sodium buildup hazard by displacing sodium in the soil. $CO_3 + HCO_3$ increase sodium buildup hazard by precipitating calcium and magnesium as carbonates which have an extremely low solubility.

Four different methods of assessing the potential hazard of soil sodium buildup (or soil permeability problems) are in general use. These 4 methods are discussed below (**note:** in all calculations given below, all concentrations of chemical constituents are given in meq/l):

1. **Sodium Adsorbtion Ratio (SAR)**

$$SAR = \frac{Na}{\sqrt{Ca+Mg/2}}$$

SAR was one of the first methods of assessing the potential hazard from sodium buildup but has given way to other methods that consider the $CO_3 + HCO_3$ content. In general, waters with SAR values above 6.0 were considered to represent a sodium hazard, although at a given SAR, low salinity waters were considered much less of a sodium hazard than high salinity waters.

2. **Residual Sodium Carbonate (Res. Na_2CO_3 or RSC)**

$$RSC = (CO_3 + HCO_3) - (Ca+Mg)$$

Positive values indicate a sodium buildup hazard. Multiplying RSC by 234 gives a gypsum requirement in lbs. of 100% gypsum/acre ft. of water. (234 is the factor used to convert a calcium deficit or excess from meq/l to lbs. of gypsum per acre foot of water; if metric units are desired, the factor is 86 and the units are kg of gypsum per 1000 m3).

RSC was introduced by USDA worker Frank Eaton in 1950 in a paper that has been correctly termed a classic by his peers. Since that time, RSC has become a popular method of assessing sodium hazard and is still used today. Although RSC does not take into account the sodium content of water, it is a simple and usually accurate method of determining potential permeability problems from sodium.

3. **Eaton's Gypsum Requirement**

   Not satisfied with RSC, Eaton introduced the following method of determining the calcium deficit (or gypsum requirement) of water in 1966:

   $$Gyp.\ Req. = 234\ (a + b + c),\ where$$

   $$a = 0.43\ Na - (Ca+Mg)$$
   $$b = 0.7\ (CO_3 + HCO_3)$$
   $$c = 0.3$$

   Positive values indicate a potential sodium buildup hazard. Negative values indicate that the water can deposit gypsum.

   Eaton's Gypsum Requirement was the first method of assessing sodium hazard that considered all 3 of the influencing factors: Na, $Ca+Mg$ and $CO_3 + HCO_3$. Although not widely used, Eaton's Gyp. Req. is relatively easy to calculate and has proven reliable.

4. **Adjusted Sodium Adsorption Ratio (adj. SAR)**

   Introduced in 1972, adj. SAR uses pH_c values to modify the SAR to account for the effects of $CO_3 + HCO_3$ (see following page for determination of pH_c):

   $$adj.\ SAR = SAR\ (9.4 - pH_c)$$

   Soil permeability problems from sodium buildup are likely with adj. SAR values over 6.0. Adj. SAR is being incorporated in many new water quality guidelines and will probably be widely used in the future.

8

- CALCULATING pH_c -

pH_c values are used mainly to calculate adj. SAR values and consider the concentration of Na, Ca+Mg and CO_3+HCO_3. pH_c values will also tell whether a given water will dissolve lime or deposit lime as it passses through the soil.

$$pH_c = a + b + c, \text{ where}$$

$a = pK_2' - pK_c'$: (calculated from Ca, Mg, Na)
$b = p(Ca+Mg)$
$c = p(CO_3+HCO_3)$

a, b and c Values for pHc Formula

meq/l	- a -	meq/l	- b -	meq/l	- c -
Ca + Mg + Na	pK'2 - pK'c	Ca + Mg	p(Ca + Mg)	CO_3 + HCO_3	p(CO_3 + HCO_3)
.5	2.11	.05	4.60	.05	4.30
.7	2.12	.10	4.30	.05	4.30
.9	2.13	.15	4.12	.15	3.82
1.2	2.14	.20	4.00	.20	3.70
1.6	2.15	.25	3.90	.25	3.60
1.9	2.16	.32	3.80	.31	3.51
2.4	2.17	.39	3.70	.40	3.40
2.8	2.18	.50	3.60	.50	3.30
3.3	2.19	.63	3.50	.63	3.20
3.9	2.20	.79	3.40	.79	3.10
4.5	2.21	1.0	3.30	.99	3.00
5.1	2.22	1.25	3.20	1.25	2.90
5.8	2.23	1.58	3.10	1.57	2.80
6.6	2.24	1.98	3.00	1.98	2.70
7.4	2.25	2.49	2.90	2.49	2.60
8.3	2.26	3.14	2.80	3.13	2.50
9.2	2.27	3.90	2.70	4.0	2.40
11	2.28	4.97	2.60	5.0	2.30
13	2.30	6.30	2.50	6.3	2.20
14	2.32	7.90	2.40	7.9	2.10
18	2.34	10.00	2.30	9.9	2.00
22	2.36	12.50	2.20	12.5	1.90
25	2.38	15.80	2.10	15.7	1.80
29	2.40	19.80	2.00	19.8	1.70

pH$_C$ values above 8.4 indicate a tendency to dissolve lime from soil through which the water moves; values below 8.4 indicate a tendency to precipitate lime from waters applied.

pH$_C$ values are helpful in assessing potential clogging problems from calcium carbonate precipitation on drip irrigation systems and in predicting scaling problems (from lime deposition) on pump bowls. An older method of assessing lime deposition, the Langlier index, is easier to calculate and is still used by some labs.

Evaluating the 4 Sodium Permeability Evaluations

Although **SAR** does not consider $CO_3 + HCO_3$, it can be useful in cases where other methods give excessive weight to $CO_3 + HCO_3$ (see following section).

adj. SAR will probably become the accepted standard by which to assess potential sodium permeability problems. It has the disadvantage of not being easily calculated (I have seen several calculation errors by labs, so it pays to verify a lab's calculation) and not being easy to explain to a grower.

Residual sodium carbonate is easily calculated, but because it does not consider sodium content, there will be a few cases where other methods are superior.

I have made extensive use of **Eaton's Gyp. Req.** and have found it to be a useful and reliable tool. It also gives a good handle on the amount of gypsum deposited by a water - useful when making gypsum recommendations.

A comparison of the 4 methods is shown in the following.

Water Examples

Analysis of 8 waters is given in Table 2 along with observed permeability problems, if any, associated with each.

Table 2. Water Analysis Data for Eight Different Waters*

No.	Description	Permeability Problems?	Year Sampled	pH	EC mmhos /cm	Ca+Mg	Na	CO3+ HCO3	Cl	SO4	pHc	SAR	adj. SAR	Resid. Na2CO3	Eaton's Gyp Req.**
								meq/l					Permeability Evaluation		
	CALIFORNIA WELLS														
1.	A	Yes	1971	7.8	0.38	1.00	2.46	0.69	1.5	1.2	8.7	3.5	2.6	0	+197
2.	B	Yes	1975	9.3	0.48	0.94	3.87	0.92	1.7	2.0	8.6	5.7	4.6	0	+390
3.	C	No	1978	7.5	1.24	5.72	6.61	2.66	8.9	0.4	7.4	3.9	7.7	0	-168
4.	D	Yes	1977	7.9	0.32	0.56	2.74	2.70	0.4	0.2	7.7	5.2	5.6	2.14	+657
	RIVERS														
5.	Colorado	No	1943	•	1.06	6.90	4.06	2.64	2.1	6.4	7.3	2.2	4.6	0	-702
6.	Pecos	•	1946	•	3.21	26.48	11.52	3.18	12.0	23.1	6.7	3.2	8.5	0	-4445
	SEWAGE WATER														
7.	Morro Bay effluent	No	1976	7.6	2.0	6.79	11.69	8.28	9.0	1.7	6.9	6.4	15.9	1.49	+1015
	WASHINGTON WELL														
8.	North central area	•	1944	7.8	0.64	5.32	1.31	5.03	0.1	1.8	7.1	0.8	1.8	0	-220

* information not available
** lbs/ac. ft. of water. negative values indicate gypsum deposited
*** Source of data for the 8 waters is given below

1-4 Personal data from files

5 USDA Handbook No. 60. **Diagnosis and Improvement of Saline and Alkaki Soils (1954)**

6 FAO Irrigation and Drainage Paper No 29. **Water Quality for Agriculture.** R S Ayers and D W Westcot (1976)

7 Wildman, W E, R L Branson, J M Rible and W E Cawelti. **Irrigation trial with Morro Bay wastewater.** Calif. Agric. May 1977, pp 36-37

8 Harley, C.P. and R C Lindner. **Observed responses of apple and pear trees to some irrigation waters of north central Washington.** Proc Amer Soc Hort Sci 46 35-44. (1945).

11

Permeability problems on Nos. 1, 2 and 4 are due to low salt content as well as to gypsum requirement.

The adj. SAR value of 8.5 for No. 6 (Pecos River) indicates a potential permeability problem due to sodium, however with the water supplying over 2 tons of gypsum per acre foot, one should be safe in predicting that permeability problems will not occur. The sodium chloride content of this water will present problems unless careful management is used.

Eaton's Gyp. Req. shows that the Colorado River is an excellent source of gypsum. This river has received a bad reputation in the past because of it's relatively high total salt content, but well over half of the salt content is gypsum - a benign salt for agriculture.

Morro Bay Effluent (No. 7)

From looking at the Permeability Evaluation of the Morro Bay effluent (No. 7), one would predict permeability problems from sodium buildup. Extensive tests with this water, however, showed that soil SAR values leveled off at 4.5 to 5.5 after 5 feet of water had been applied and did not increase when up to 17 ft of water had been applied. What happened?

The logical explanation is that the soil was kept wet enough so that calcium carbonate did not have a chance to precipitate. The entire 17 feet of water was applied within a year - this in a cool climate. The permeability evaluations by adj. SAR and Eaton's Gyp. Req. are based on lime (calcium carbonate) precipitating out as the soil dries. With the large amounts of water applied, much of the bicarbonate would be leached below the root zone and thus not have a chance to react with calcium to precipitate lime. Supporting this explanation is the fact that the soil SAR actually **decreased** after the water application rate was doubled during the course of the experiment.

The preceding explanation is in agreement with a study that measured lime precipitated from irrigation water applied to a California citrus grove over a 38 year period. The amount of lime precipitated was less than expected in 2 of the treatments and the authors concluded that "The most logical explanation for the low amounts of $CaCO_3$ precipitated is that the soil was well leached."

12

Both the Morro Bay effluent data and the 38 year citrus grove experiment indicate that with a well drained soil and high water application rates, the harmful permeability effects of $CO_3 + HCO_3$ are greatly reduced. The same would be expected to hold true for drip irrigation, where soils are kept on the wet side. Under such conditions, the SAR values (which consider only $Ca + Mg$ and Na) would better predict permeability effects; this was indeed the case with the Morro Bay effluent.

The Morro Bay effluent data show that rigid interpretation of water analysis data should be avoided, particularily when $CO_3 + HCO_3$ are involved. Eaton recognized this fact when introducing his Gypsum Requirement formula: "The factor here suggested for bicarbonate precipitation (HCO_3 X 0.7) may fit some situations well enough, but frequent and possibly wide departures will inevitably occur."

Washington State Well (No. 8)
If you took the water analysis data for sample No. 8 to 10 different water quality experts, at least 9 would say that the water quality was excellent for agriculture and that no problems should be anticipated other than possible scale buildup on pump bowls. The study from which this water data was taken states, however, that apple and pear trees irrigated with such waters over a period of years show "definite manifestations of vigor decline accompanied by varying degrees of chlorosis". Examination of tree roots showed lime incrustations around many of the roots and the decline and chlorosis was attributed to these incrustations.

The pH_c value and the $Ca + Mg$ and $CO_3 + HCO_3$ content would predict lime deposition, but would not predict any problems from such deposition. Perhaps roots of some plant species encourage lime precipitation around the root itself.

The Washington State study also warns against rigid interpretation of water analysis data. Available criteria are guidelines only - the mind must always be kept open.

Increased Hazards from Sprinkler Irrigation
Leaves absorb both sodium and chloride, therefore sprinkler irrigation can present a hazard to sensitive crops. For sprinkler

irrigation, both sodium and chloride should be less than 3 meq/l; when sprinklers are used for crop cooling (frequent on-off cycles), sodium and chloride should be less than 2 meq/l.

If leaf burn does occur when using sprinklers, irrigate at night; avoid irrigating during the hottest part of the day or during windy periods as conditions conducive to high evaporation will increase leaf burn. Increasing the speed of rotation of sprinkler heads will reduce leaf burn hazard by not allowing salts to concentrate (through evaporation) in water that has been applied.

Bicarbonate contents in excess of 1.5 meq/l can give white deposits (mainly lime) on leaves and fruit. Again, night sprinkling will minimize this problem (if it is a problem). Neutralizing the bicarbonate by the addition of acid to the water is a satisfactory alternative.

Water Ammendments

As indicated previously, nothing can be done to change the quality of a water that has an excess of the 3 main toxic constituents, however when water quality is impaired because it has a potential permeability hazard from sodium buildup, the quality of the water can be improved by the addition of ammendments.

A look at the formulas used to calculate permeability hazards shows that a reduction in permeability hazard will occur by increasing the Ca content or by reducing the HCO_3 content. Addition of gypsum (calcium sulfate) is a common method of improving the Ca content of waters. Addition of sulfuric acid or SO_2 will reduce the HCO_3 content. Ammendments do not have to be added directly to the water, but can be applied to the soil on which the water is to be used.

Chapter 4 discusses ammendments in more detail.

Long Term Effects of Water on Soils

It is written (or some day will be): "as the water is, so then shall be the soil." If a given soil is irrigated with a given water over a long period of time, the soil will assume the characteristics of the water. In such cases, a look at water analysis data can provide an insight into the soil, even if no soil analysis data are available. Conversely, soil analysis data can indicate the chemical composition of the irrigation water used.

An interesting corollary to the above is water's effect on soil pH. The acid soils of the world invariably occur in high rainfall areas. Rain water, which has a high pH_c has acidified the soil.

Much of the irrigation water used on the east side of the San Joaquin Valley in California is runoff from the Sierras. This water, not having traveled far, is relatively pure and has a high pH_c. In effect, by using transported rainwater from the Sierras, some land on the eastern side of the San Joaquin Valley has been turned into the equivalent of a high rainfall area.

How long does it take for a high pH_c water to acidify a soil? A rough approximation can be made by using the calculations used to derive pH_c. Assuming the following water content:

$Ca+Mg$ = 0.62 meq/l; Na = 0.20 meq/l; $CO_3 + HCO_3$ = 0.52 meq/l, then

pH_c = 8.92 where a = 2.12, b = 3.50 and c = 3.29 (see table for calculating pH_c)

then,

1. Determine what value factor "c" should have in order to make the water a neutral water (pH_c of 8.4); "c" should be 2.77 in order to give a pH_c of 8.4 (a + b + 2.77 = 8.4).

2. From the pH_c table, determine the $CO_3 + HCO_3$ content needed to give a "c" value of 2.77. The answer is 1.69 meq/l.

3. Determine the difference between the $CO_3 + HCO_3$ content needed to give a neutral water and the actual content:

$1.69 - 0.52 = 1.17$ meq/l of HCO_3 difference

This is the bicarbonate deficit of the water (compared to a neutral water having a pH_c of 8.4).

It takes roughly 135 lbs of 98% sulfuric acid per acre foot of water to neutralize the bicarbonate represented by 1 meq/l. Therefore, an acre foot of water that has a bicarbonate deficit of 1.17 meq/l has the acidifying potential of 158 lbs of sulfuric acid (135 X 1.17). Using 3 acre feet of this water each year can be the same as adding 474 lbs of sulfuric acid per acre each year and can cause significant soil acidification over a period of years if the soil does not contain lime*.

Over a 10 to 20 year period, the water with a pH_C of 8.92 can have a significant effect on the pH of coarse textured soils that do not contain lime. Under a drip irrigation system that wets 1/4 of the total soil, the acidifying effects will take place 4 times as fast.

In the case of a water that deposits lime (pH_C less than 8.4) it can similarily be determined how much lime the water will deposit per acre foot used by calculating the bicarbonate excess (rather than deficit) of the water (i.e., the amount of bicarbonate in excess of that needed to give a pH_C of 8.4). For each meq/l of HCO_3 excess, the water will deposit 135 lbs. of lime per acre foot of water used. A lime depositing water will tend to stabilize soil pH in the 7.2 to 7.7 range (and will gradually raise the pH of an acid soil to this range).

Some fieldmen are rightfully concerned about the acidifying effect of nitrogen fertilizers on soils. On non-calcareous soils, continued application of acidifying N fertilzers will lower soil pH - possibly enough to be detrimental to crop growth.

On a soil that contains 40,000 lbs of lime per acre foot (as some soils do) the acidifying effects of N fertilizers are of no immediate concern. On soils that do not contain lime, the acidifying effects of most N fertilizers can be easily offset if the irrigation water has a low pH_C. (see chapter on nitrogen for a further discussion of acidifying effects of N fertilizers).

---

* if the soil has a high lime content, the amount of acid will have no effect on soil pH until the lime is neutralized (soil pH stays in the 7.2 to 7.7 range as long as there is significant lime in the soil). The soil series (from a soil map) will tell whether a given top soil contains lime in it's virgin state (use of low pH_C water over a period of years can add lime to a lime free soil). A lab test can also be used.

Acidifying N fertilizers in combination with high pH_c irrigation water will eventually cause excessive soil acidity on soils that do not contain lime in the top soil. This combination has caused excessively acid soils on the east side of California's San Joaquin Valley. Use of elemental sulfur in pest and disease control programs will compound such acidity problems.

It should be remembered that pH_c values indicate a *tendency* to dissolve or precipitate lime. Calculations meant to predict a water's long term effect on soil pH are rough estimates only and are not meant to be precise.

Water as a Source of Nutrients

House plants irrigated with tap water thrive - the same plants irrigated with distilled water can do poorly. Why?

Obviously there are nutrients in tap water that plants can use. All water sources pass through mineral strata prior to being tapped for irrigation. Mineral content of water varies, depending on where the water comes from and where it's been.

The amount of nutrients supplied by an acre foot of a given water source can easily be calculated from water analysis data. First determine the amount of the nutrient in ppm, then multiply by 2.7 (an acre foot of water weighs 2.7 million lbs.) to get lbs per acre foot.

Most water sources in the western U.S. contain sufficient magnesium and sulfur to supply a crop's needs. Crop response to sulfur in the western U.S. virtually always occurs when the crop is dependent solely on rainfall for it's water or when irrigation is from a water source with a very low sulfate content. As little as 5 lbs of S per acre foot of water will supply the sulfur needs of most crops. This equivalent to a SO_4 content of 5.5 ppm or 0.12 meq/l.

Of the 3 major nutrients, water is rarely a significant source of potassium or phosphorus but some waters have a high enough nitrate content to supply a good portion of a crop's N needs. Needless to say, this can be a drawback for crops that are sensitive to excess N.

Comparing Amounts of Sodium and Chloride Added by Water to Amounts Added by Fertilizers

Table 1 shows that a good water (containing 6.5 meq/l Na and 4.2 meq/l Cl) can supply 400 lbs. of sodium and 400 lbs. of chloride per acre foot of water used. Assuming 3.5 acre feet of water are used during the growing season, over 1400 lbs each of sodium and chloride would be added to the soil by such a water. Table 1 says that most crops can take the addition of 1400 lbs of sodium or chloride during a growing season with no undue concern about adverse effects.

There are 2 fertilizer materials that contain significant amounts of sodium and chloride: sodium nitrate (or nitrate of soda), a nitrogen fertilizer, and potassium chloride (or muriate of potash), a potassium fertilizer. Many fieldmen are leery of recommending these materials because of their sodium or chloride content, although muriate of potash is much more economical than other potash sources and there are instances (on acid soils, or where a quick N boost is desired) where sodium nitrate might be practical.

To determine if such materials are safe, determine the amount of sodium or chloride supplied by the material and add to this figure the amount supplied by the irrigation water during the growing season. If the total is less than 1400 lbs, the material can be safely broadcast (for very sensitive crops a 700 lb limit should be used; for crops such as avocados and strawberries, that are extremely sensitive to chloride, no chloride should be applied).

Examples:
1. Is it safe to broadcast 100 units of N as sodium nitrate to a crop of corn if the water supply contains 2 meq/l Na? (assume 3.5 acre feet of water used during the growing season).

   a) Na supplied by water during season = 435 lbs (2 X 23 X 2.7 X 3.5)

   b) Na supplied by sodium nitrate = 168 lbs (sodium nitrate is 16% N, 27% Na; 625 lbs. of material supply 100 lbs of N and 168 lbs of Na)

18

c) total sodium added = 603 lbs.

Answer: Why not?

    Question: How would the sodium content from the above material effect water penetration?

        Answer: Not much. If you spread the 168 lbs of sodium over 3.5 acre feet of water you increase the effective sodium content by 17.7 ppm or 0.77 meq/l. This amount of sodium increases Eaton's Gyp. Req. by +77 lbs/acre foot of water (0.43 X 0.77 X 234). If water penetration was a concern, gypsum could be added. Water penetration for the initial irrigation might be poorer, but should improve during subsequent irrigations.

2. Is it safe to broadcast 200 lbs of K_2O as muriate of potash to a crop of tomatoes if the irrigation water contains 1.5 meq/l Cl? (assume 3.5 acre feet of water will be applied during the growing season).

    a) Cl supplied by water during season = 503 lbs

    b) Cl supplied by muriate of potash = 170 lbs
    (muriate of potash is 61% K_2O; 328 lbs supplies 200 lbs K_2O and 170 lbs of Cl).

    c) total chloride added = 673 lbs

Answer: Yes

It should be noted that the above examples specify broadcast application. Band application of the same materials would be too hazardous for many crops, especially in the seedling stage.

General References

1. Ayers, R.S. and D.W. Westcot, **Water Quality for Agriculture,** Irrigation and Drainage Paper No. 29, Food and Agriculture Organization of the United Nations (FAO), Rome 1976.

2. Eaton, Frank M., **Significance of carbonates in irrigation water,** Soil Sci. 69:123-134 (1950).

3. Eaton, Frank M., **Total salt and water quality appraisal,** pp. 501-509, in Diagnostic Criteria for Plants & Soils, Homer Chapman Ed., University of California Div. of Agric. Sciences, (1966).

4. Stromberg, L.K., **Water Quality for Irrigation,** Fresno County Ag Extension Service, 1720 S. Maple Ave., Fresno, Ca. 93702. (1975).

5. USDA Handbook No. 60, **Diagnosis and Improvement of Saline and Alkali Soils** (1954). (A revised edition should be available in 1982).

2. BICARBONATE
The Elusive One

Dave West* has aptly termed bicarbonate a "will-o'-the-wisp". The "now you see it, now you don't" nature of bicarbonate makes it difficult to secure a handle on it.

The bicarbonate concentration of the soil solution varies with the moisture content of the soil, the lime, gypsum and calcium content and the phosphorus and iron content. The CO_2 producing capacity of plant roots also influences soil bicarbonate concentration and there are other influencing factors.

Although bicarbonate is routinely run in water analysis, it is not routinely run on the saturation extract of soils - it should be, as the analysis is easy and could provide useful information. Perusal of accumulated bicarbonate analysis data from large numbers of soil samples could provide interesting correlations and aid in diagnosing field problems.

Bicarbonate concentration of plant cell sap is probably a critical factor in bicarbonate related problems, but how do you get a handle on this? Plant analysis for bicarbonate is virtually never done because of the difficulty of correlating results with the actual cell sap concentration. Virtually any type of processing of plant tissue (drying, ashing) will immediately alter the bicarbonate concentration. Perhaps an extract of fresh tissue could provide clues, but this is never done.

Evidence that bicarbonate can be directly toxic to plants was shown in an experiment with radish seedlings by comparing equal concentrations of sodium chloride (NaCl) and sodium bicarbonate ($NaHCO_3$):

RELATIVE ELONGATION OF RADISH
RADICLES IN TWO DIFFERENT SOLUTIONS

Solution	0	5	10	50	100	200
NaCl	100	94	86	81	75	60
$NaHCO_3$	100	90	78	36	18	2

-Solution concentration in meq/l-

*formerly Kern County (Ca) farm advisor, now in private consulting

21

Chloride is usually considered the most toxic anion to plants. The preceding data indicate that bicarbonate is more toxic to some species.

Bicarbonate is strongly implicated in iron chlorosis (see chapter on iron). One proposed mechanism for the relationship is that bicarbonate makes phosphorus more available by tying up calcium and thus increasing the solubility of calcium phosphates (remember that a bicarbonate extraction is a standard laboratory method for soil phosphorus analysis). High phosphorus levels in plant tissue are often associated with iron deficiency and can sometimes be used as the sole diagnostic confirmation of an iron deficiency.

Until the day that bicarbonate guidelines for soil and water become better established, the fieldman should pay close attention to water **and** soil bicarbonate levels when diagnosing field problems, particularly chlorosis problems.

GENERAL REFERENCE
The May 1960 issue of **Soil Science** (Vol. 89, No. 5) consists of 11 papers on bicarbonate that were presented as a Symposium on Bicarbonates in 1959.

3. SALINE AND ALKALI SOILS

There is probably more excellent, reliable, useful information on saline and alkali soils than on any other phase of soil and plant nutrition. This is due mainly to the efforts of a dedicated band of USDA scientists who, in the nineteen forties and early fifties put out a wealth of solid information on the subject. Much of this information is compiled in USDA Handbook 60, published in 1954 (see General Reference at end of chapter). A revised version of this publication is scheduled to be released by 1982 and this revised version should be part of every fieldman's library.

An excellent, concise, publication on the reclamation of saline and alkali soils was published by Stromberg in 1972 (see General Reference).

The classes of saline and alkali soils are given below:

CLASSIFICATION OF SALINE & ALKALI SOILS*

Soil Class	EC	ESP	Recommended Reclamation
Saline	Over 4	Under 15	Drainage and leaching
Saline-alkali	Over 4	Over 15	Leaching followed by ammendment if ESP stays over 15
Nonsaline-alkali	Under 4	Over 15	Ammendment + leaching
Normal	Under 4	Under 15	—

It should be noted that pH is not a criterion used to designate alkali soils and that alkali soils can have an acid pH (below 7.0) although this is rare; soils can also have an alkaline pH (above 7.0) but not be alkali soils. Alkali soils would be more properly termed sodium soils or sodic soils, and in some cases they are. The confusion between alkaline, alkali and alkalinity will probably always be with us.

Soil pH values in excess of 8.5 almost invariably indicate an alkali soil (ESP over 15). This fact has been used for field classification of alkali soils. Phenolphtalein indicator changes color in the 8.5 to

---

*EC given in mmhos/cm; ESP = exchangeable sodium percentage

9.0 range and a positive phenolphtalein test will therefore indicate an alkali soil. Many soils in California have been mapped as alkali soils based on the phenolphtalein test.

Soils can be alkali, however, and have a pH less than 8.5. Saline-alkali soils **usually** have a pH reading below 8.5. Unfortunately some of the old soil maps classify some soils as non-alkali because they didn't give a positive phenolphtalein test; many of these soils are saline-alkali - the salts in the soil modify the pH to a lower range. Phenolphtalein essentially delineates only **non-**saline alkali soils.

Also useful when interpreting soil analysis data is that EC values of soils containing gypsum should be reduced by 2 mmhos/cm for interpretive purposes. On the other hand, EC levels of very sandy soils (SP below 30) should be multiplied by 2 before interpreting and EC values of moderately sandy soils (SP between 30 and 40) should be multiplied by 1.5 before interpreting. In other words, a given EC reading is more hazardous the sandier the soil.

The following reclamation schedule is modified from Stromberg (see Gen. Ref. 2) and can be used as a guide in reclaiming saline-alkali soils:

SAMPLE RECLAMATION SCHEDULE
FOR SALINE-ALKALI GROUND

Summer Months (1st year)
1. Grade the land for surface irrigation.
2. Rip to a depth of 30'' or more; shank spacing on crossbar should be no greater than the depth of ripping.
3. Disc to smooth surface and make some temporary irrigation borders.
4. Irrigate the ground at least once or twice until the soil seals up.
5. Apply 10 tons of 55% gypsum* per acre.

---

* ammendment choice is not limited to gypsum. 1 ton of elemental sulfur is equivalent to 10 tons of 55% gypsum and can be substituted for gypsum on most soils. Sulfuric acid has also proven to be an excellent ammendment on some soils. (see separate chapter on ammendments for further information).

6. Disc and cross disc the gypsum to mix it in the top 6 inches of soil.
7. Irrigate again.

Fall Months (1st year)
1. Work up a seedbed.
2. Apply 80 units of nitrogen per acre.
3. Seed a winter cereal, preferably barley.

Winter Months (1st year)
1. Irrigate during winter and early spring to keep crop growing and to leach soil.
2. Map alkali areas (poor growth; slick spots) for additional gypsum application.

Spring and early Summer Months (2nd year)
1. Harvest cereal.
2. Treat areas mapped (above) and any bare spots in the fields with another 10 tons of 55% gypsum per acre.
3. Apply 20 gals. of 10% zinc sulfate solution per acre if soil analysis shows low zinc levels.
4. Disc stubble and mix in the gypsum (and zinc, if applied).
5. Irrigate and work up a flat seedbed.
6. Apply 100 units of nitrogen per acre.
7. Seed to sudangrass or grain sorghum.
8. Irrigate by flooding between temporary borders. Avoid furrow irrigation because this concentrates the released salts in the tops of the beds.

Late Summer and Fall Months (2nd year)
1. Harvest the crop.
2. Use soil analysis to determine if additional gypsum should be applied.
3. Apply gypsum if soil tests show need, using twice as much on bare or weak spots.
4. Plant to irrigated pasture or alfalfa. If the process of reclamation has been good, an alternative may be to plant the land to a fall seeded row crop or to a spring seeded row crop such as corn or cotton. Once salinity levels have been lowered to a safe range, furrow irrigation can be used.

If it is known that a particular piece of ground is saline or saline-alkali (and it usually is) extensive soil sampling prior to the first irrigation is not warranted since once the first irrigation has been applied the entire chemical composition of the soil is drastically altered. For example, 1 acre foot of water will remove approx. 80% of the salts from a 1 foot depth of soil. Why spend alot of money on initial soil sampling when the results will be meaningless after the first irrigation? Soil analysis of more value *after* the initial leaching; this is the analysis that can determine the amounts of ammendments needed, if any.

As indicated in the preceding reclamation schedule, particular attention should be given to mapping and reclaiming the weak spots that will always occur in a parcel of any size. These spots can be mapped visually; there is no need in running extensive soil tests except to substantiate an opinion on a particular area. An aerial photo during the growing season can be much more useful than soil analysis in mapping these weak areas - I have made extensive use of such photos; nothing fancy is needed, a plain old black and white photo will suffice. It makes no sense to apply ammendments to an entire field when only a relatively small portion will reap the benefits - in other words, put your money where it will do the most good.

General References
1. Richards, L.A. (ed.) **Diagnosis and improvement of saline and alkali soils.** USDA Handbook 60. 1954.

2. Stromberg, L.K. **Reclaiming saline and alkali soils.** 8 page mimeo from the Fresno County Ag Extension Service, 1720 S. Maple, Fresno, CA 93702. 1972.

3. See chapter on ammendments, following.

4. AMMENDMENTS

Ammendments are used for 2 main reasons:

1. **Reduction of soil alkalinity**
   Soils with ESP values greater than 15 are classified as alkali soils and ammendments should be applied to such soils. For grapes and tree crops, ESP levels should be below 10 (probably below 5 for some species) and ammendments should be applied to maintain ESP below these levels.

2. **Improvement in water penetration**
   Impaired water penetration can be caused by low salt water or by water having an unfavorable ratio of sodium and bicarbonate to calcium and magnesium (see Chapter 1). It must be remembered that poor water penetration can be caused by compacted soil, restrictive soil layers and soil crusting, all of which will be better remedied by mechanical means (ripping, harrowing) than by ammendments. An actual need for ammendments should exist before they are applied.

Ammendments exert their benefit by increasing soluble calcium levels in the soil. Soluble calcium replaces sodium on the exchange complex and provides for a looser, more friable soil (sodium tightens soils, calcium loosens soils). A reduction of sodium, by definition, means a reduction in soil alkalinity. Ammendments fall into 2 broad categories.

A. **Ammendments containing calcium directly; this includes:**
   1. **Gypsum** (calcium sulfate) the most widely used soil ammendment.

   2. **Lime** (calcium carbonate) which is only an effective ammendment if the soil is acid, since lime is only soluble in acid soils.

   3. **Calcium polysulfide** or lime sulfur - primarily an acidifying ammendment (see following).

27

B. **Acidifying ammendments that release calcium from lime (calcium carbonate)**

In order for acidifying ammendments to be effective, there must be lime in the soil. Acidifying ammendments include:

1. **Sulfur** - sulfur is biologically converted in the soil to sulfuric acid; 1 ton of sulfur produces approximately 3.2 tons of sulfuric acid.

2. **Sulfuric acid**

3. **Sulfur dioxide (SO_2)** - usually added by burning sulfur to release SO_2 which is mixed with water to form sulfurous and sulfuric acid.

4. **Polysulfides** - the polysulfides are liquids and contain un-oxidized sulfur in the sulfide form. This sulfur is biologically converted to sulfuric acid. There are 2 polysulfide ammendments in commercial use:

   a) **Calcium polysulfide (CaS_5),** also called lime sulfur. 100 lbs of this material will produce 68 lbs. of sulfuric acid. The calcium in the material is of direct benefit.

   b) **Ammonium polysulfide ($(NH_4)_2S_x$** - also called Nitro-Sul. The ammonium in the material also acts as an acidifying ammendment (as well as a nitrogen fertilizer); 100 lbs of material will produce the equivalent of about 240 lbs of sulfuric acid.

5. **Ammonium Thiosulfate ($(NH_4)_2S_2O_3$** - also called Thio-Sul (a liquid). The partially oxidized thiosulfate form of sulfur is biologically converted to sulfuric acid. The ammonium also acts as an acidifying ammendment. 100 lbs of material will produce the equivalent of 102 lbs. of sulfuric acid.

Choice of Ammendments for
Reducing Soil Alkalinity

For reducing soil alkalinity, large amounts of ammendments are needed, and either gypsum or sulfur, or in some cases sulfuric acid is used. A comparism of these 3 materials is given below:

Ammendment	Tons required to equal 10 tons of 60% gypsum
60% gypsum	10
Sulfur	1.1
Sulfuric acid	3.5

To determine the most economical ammendment to apply, the costs of each must be known. Gypsum and sulfur are easily more economical than sulfuric acid on an equivalent cost basis, however sulfuric acid can out perform other ammendments in the field. Gypsum is the most widely used ammendment. Sulfur has the disadvantage of not being immediately active since it must undergo biological conversion to sulfuric acid. The rate of conversion depends on soil moisture and temperature and could take anywhere from a few weeks to over a year.

Sulfuric acid, although more expensive, has given excellent results in many situations and in test situations it has out performed both gypsum and sulfur on an ammendment equivalent basis. Sulfuric acid goes to work immediately and it may be that its quick action on lime results in a larger percentage of lime being converted to gypsum (bicarbonate intermediates may not be allowed to react with the acid). Sulfuric acid has the disadvantage of being a hazardous material to apply.

Choice of Ammendments for
Improving Water Penetration

For the improvement of water penetration it should be kept in mind that many acid soils can have water penetration problems and that acidifying ammendments are of no value on acid soils. Acidifying ammendments are only effective if there is lime in the soil and can actually aggravate penetration problems on acid soils.

Lime is an excellent and economical ammendment choice for improving water penetration of acid soils; 1 ton of lime has the ammendment equivalent of 3 tons of 60% gypsum.

When it is desired to add ammendments to the water, the choice is pretty well limited to the liquid acidifying ammendments and SO_2 because of the low solubility of gypsum (and the virtual insolubility of lime). Gypsum can be added to water but it must be metered slowly and mixed thoroughly; few people are successfully adding gypsum to water.

When adding acidifying ammendments to water, care should be taken not to allow the pH to drop much below 6.5 to prevent damage to irrigation lines and equipment. Neutralizing half the bicarbonate in a given water will usually drop the pH to the 6.5 range. It takes roughly 135 lbs of sulfuric acid (or 45 lbs of sulfur, burned to give SO_2) per acre foot of water to reduce the bicarbonate (or carbonate) concentration of a given water by 1 meq/l. For each meq/l that the bicarbonate content is lowered, Eatons Gyp. Req. (see Chapter 1) is reduced by 164 lbs.

Other ammendments (e.g., lime sulfur) can have a corrosion hazard independent of acidity. Before running a given ammendment through an irrigation system a knowledge of it's corrosion hazard should be obtained.

When water penetration problems occur on an acid soil, it is unwise to add acidifying ammendments to the water **unless** lime is first added to the soil. This combination - lime on the soil, acidifying ammendments in the water - has been successfully used on acid soils and will probably see more use in the future. On neutral soils irrigated with a water that has a high pH_c (dissolves lime), lime is also a satisfactory ammendment choice (for soil application).

The simultaneous addition of NH_3 and sulfuric acid to irrigation water has been successfully used in Arizona. Essentially ammonium sulfate is formed and the corrosion hazard of the acid is neutralized by the NH_3; volatilization loss of NH_3 is minimized or eliminated by the sulfuric acid. Proper metering and placement is important when using NH_3 with sulfuric acid but the method holds promise for those that attempt it in a diligent manner.

SO_2 generators have gained widespread popularity in the southern San Joaquin Valley of California in recent years and have been effective in improving water penetration. With both SO_2 and sulfuric acid, the amount that can be safely added (without excessive acidification of the water) is determined by the bicarbonate content of the water source - the higher the bicarbonate content, the more acid can be safely added.

Calcium nitrate - combination fertilizer and ammendment

The ammendment properties of calcium nitrate should be kept in mind when planning a fertilizer program for a crop on which water penetration is problem. Calcium nitrate contains 20% calcium and is an extremely soluble material. It is approximately 500 times more soluble than gypsum and because of this high solubility, 1 lb of calcium from calcium nitrate is probably worth more than a pound of calcium from gypsum. Calcium nitrate is another weapon in the ammendment arsenal of the fieldman. (Calcium ammonium nitrate, a liquid, is also available and can be a useful material).

Chemicals used in pest, disease and weed control that can also be ammendments

There are three sulfur materials used in pest, disease and weed control that can also act as ammendments. They are:

1. **Elemental sulfur**
   An old standby for mite and insect control is gaining renewed interest in some areas. On grapes, sulfur has been regularly used for mildew control for years.

2. **Lime sulfur**
   A widely used material for scale control at the turn of the century and also effective as a fungicide. Still recommended and used in some areas as both an insecticide and a fungicide.

31

3. **Sulfuric acid**
   Used for weed control; can be the best choice for weed control of certain crops, such as onions. Application is hazardous but there are custom applicators that specialize in sulfuric acid application for weed control.

On calcareous soils where ammendments can be beneficial, efforts should be made to integrate the above materials into the total farm management program. Because all of the above materials are acidifying, they will have the optimum ammendment effect on calcareous soils. On non-calcarous soils they should be used with caution or used in conjunction with lime application to the soil to guard against excess soil acidity.

General References

1. Stroehlein, J.L. **pH control on alkaline soils.** Solutions, May-June 1980. p 81-91.

2. Stromberg, L.K. and S.L. Tisdale. **Treating irrigated arid-land soils with acid-forming sulphur compounds.** Technical Bulletin No. 24. March 1979. The Sulphur Institute, 1725 K St., NW, Washington, D.C. 20006.

3. Branson, R.L. and M. Fireman. **Gypsum and other suitable ammendments for soil improvement.** Univ. of Calif. Ext. Leaflet 2149, March 1980.

PART II:

THE NUTRIENTS

Major Nutrients: N, P, K, S, Ca, Mg, Fe
Minor Nutrients: Zn, B, Mn, Cu, Mo, Cl

5. NITROGEN (N) - The Big Guy

Because of its overriding importance in agriculture, nitrogen has been studied more than any other nutrient. No attempt will be made here to give a detailed picture of nitrogen nutrition; some salient points will be discussed:

Nitrogen Fertilizers

Advantages and disadvantages of some of the more common N fertilizers are given below:

Anhydrous Ammonia (NH3)   82%N
ADVANTAGES:
High %N; easy to apply; no residue; little danger of leaching.
DISADVANTAGES:
Uneven distribution and losses in irrigation water with some systems; can't be used with sprinklers; loss potential with dry injection.

Ammonia solution (aqua ammonia)   20%N
ADVANTAGES:
Easy to apply; no residue
DISADVANTAGES:
Same as for anhydrous.

Ammonium sulfate   21%N
ADVANTAGES:
Acid residue for alkaline soils; minimal leaching loss; easy to handle; S boost, if needed.
DISADVANTAGES:
Acid residue for acid soils. Delayed availability during nitrification; high loss potential on calcareous soils if not incorporated.

Ammonium nitrate   33%N
ADVANTAGES:
High %N; no residue; half available now, half later; minimal volatilization loss potential.
DISADVANTAGES:
Some volatilization loss on calcareous soil if not incorporated (much less loss potential than am. sulfate).

Ammonium phosphate-sulfate (16-20-0)   16%N
ADVANTAGES:
Same as am. sulfate; high P content where needed; S boost, if needed.
DISADVANTAGES:
Same as for am. sulfate + higher cost; relatively low analysis.

Ammonium phosphates (11-48, 18-46, 10-34)   10+%N
ADVANTAGES:
High P content where needed.
DISADVANTAGES:
Same as for am. sulfate; low %N (can be advantage in some cases).

Calcium nitrate   15½%N
ADVANTAGES:
Calcium residue good on acid and sodic soils and for calcium stress; immediately available; no volatilization losses; very soluble (for water runs).
DISADVANTAGES:
Susceptible to leaching loss; can clump in moist weather; relatively high cost per unit of N.

Urea   42%N
ADVANTAGES:
High %N; no residue.
DISADVANTAGES:
Can be toxic at high smounts in some situations; high loss potential if not incorporated or watered in.

UN-32 Solution (half am. nitrate, half urea)   32%N
ADVANTAGES:
High %N; some available now, some later; easy to handle.
DISADVANTAGES:
Same as for urea.

Sodium nitrate   16%N
ADVANTAGES:
Immediately available; no volatilization.
DISADVANTAGES:
Sodium detrimental to soils if not countered by calcium; relatively low analysis.

Significant nitrogen losses can occur from volatilization of ammonium compounds when applied to the surface of soils. These losses are greatest on calcareous soils or when soil pH is above 7.0. Incorporating ammonium-N fertilizers reduces volatilization losses. (Urea is different from other N fertilizers; it is rapidly converted to the ammonium form after application but can be watered in rather than incorporated; if it is watered in immediately after application volatilization losses will be minimal). The table below compares volatilization loss potential of different N fertilizers:

POTENTIAL VOLATILIZATION LOSS OF N FROM DIFFERENT FERTILIZER MATERIALS

H = Over 40% Loss   L = 5 to 20% Loss
M = 20 to 40% Loss   V = Less than 5% Loss

Nitrogen Source	Applied to soil Surface and not incorporated		Broadcast on soil and disced under	Applied to soil surface and watered in		Applied in irrigation water		Banded at least 4 inches below soil surface
	Soil pH below 7 (Acid)	Soil pH above 7 (Alkaline)		Soil pH below 7 (Acid)	Soil pH above 7 (Alkaline)	Soil pH below 7 (Acid)	Soil pH above 7 Alkaline)	
Aqua-NH_3	H	H	L	H	H	H	H	V
Ammonium Sulfate	L	H	L	L	H	L	H	V
Ammonium Phosphate	L	H	L	L	M	L	M	V
Ammonium Nitrate	V	L	V	V	L	V	L	V
Calcium Nitrate	V	V	V	V	V	V	V	V
Sodium Nitrate	V	V	V	V	V	V	V	V
Urea	M	H	L	V	L	V	V	V
UN 32	V	M	V	V	V	V	L	V

As the above table shows, the potential for N loss from ammonium materials is much greater at pH levels above 7.0. If lime is applied to the surface of an acid soil (as it often is) it can greatly increase the potential of N volatilization loss on that soil.

The acidifying properties of ammonium fertilizers are a concern on non-calcareous soils, since pH levels should be maintained above 6.0 for optimum growth of most crops. The acidifying properties of some N fertilizers are given in the following table:

Material	%N	Equivalent Lbs of Sulfuric Acid or Lime Produced per 100 units of N	
		sulfuric acid	calcium carbonate (lime)
Ammonium nitrate	33	188	-
Ammonium phosphate sulfate	16	550	-
Ammonium sulfate	21	524	-
Anhydrous ammonia	82	179	-
Aqua ammonia	20	180	-
Calcium nitrate	15½	-	129
Sodium nitrate	16	-	181
Urea	45	158	-
UN-32	32	178	-

In any one year, the acidfying property of any given N fertilizer is not of concern, however over a period of 20 or more years, acidifying N fertilizers can have a significant effect on soil pH. The following table shows how sulfuric acid can affect pH of various textured soils that do not contain lime:

Lbs. of sulfuric acid per acre ft. of soil to lower pH (once all lime has been neutralized)

Soil texture	from 6.5 - 5.5	from 5.5 - 4.5	from 4.5 - 3.5
Loamy sand	3000	2400	2000
Sandy loam	5400	3400	2400
Loam	7200	5000	4000
Clay loam	9600	8000	5600

Under a drip irrigation system that wets 1/4 of the total soil, acidifying effects will take place 4 times as fast with a given amount of material as under conventional irrigation.

Nitrogen-Water Relationship

The amount of nitrogen needed for a specific crop is not a constant but varies depending on how much water is applied to the crop. With the optimum amount of water applied, the crop will need and use the optimum amount of nitrogen fertilizer. If irrigation is short of crop needs, then the nitrogen requirement is reduced. If irrigation is significantly in excess of crop needs, then the nitrogen requirement is increased to compensate for leaching losses and for possible unwanted growth. If irrigation intervals are short, the crop will draw most of it's water and nitrogen from the top soil; if irrigation intervals are long, the crop can draw most of it's nitrogen from lower soil depths. Different individual crop responses to N fertilizers can usually be traced to differences in irrigation management.

When reading reports of fertilizer trials, always look for details on the irrigation program - how much water was applied, what was the irrigation interval and what was the initial moisture content of the soil? Unfortunately, this information is not always available.

Nitrogen-Temperature Relationship
Nitrification (the conversion of ammonium-N to the more usable nitrate form) is a biological process. As such, it takes place very slowly when soil temperatures are low, but increases rapidly as soils warm up. Since plants take up most of the N in the nitrate form, this is an important consideration. At 50°F there is little nitrification. The rate of nitrification can double for each 18°F increase in soil temperature between 50°F and 90°F. At 90°F, nitrate is produced (from ammonium) 3 times as fast as it is at 50°F.

Nitrogen Toxicity
Excess nitrogen can adversely delay maturity of crops and on sugar beets will cause a reduction in sugar %. Too much N on grapes can increase shattering of the berries, thus reducing yields. On pecans, too much N can induce potassium deficiency. With current emphasis on nitrate pollution control and energy conservation, every effort should be made to use N efficiently.

Soil and Plant Analysis
Soil tests for nitrate have gained favor in recent years and can provide good supplemental information when planning fertilizer programs for row and field crops. Because of the transitory nature of nitrogen and nitrogen forms in the soil, caution should be used in interpretation of soil nitrogen levels.

Plant analysis can be very useful in adjusting nitrogen fertilizer rates for crops. N in plant tissue varies during the growing season so it is important to take samples at the proper time, where standardized reference values exist. Ideally, crops should be sampled 2 or more times a year for N analysis so that a better handle can be obtained on N nutrition for a particular crop.

Nitrification Inhibitors

There are a number of nitrification inhibitors on the market, the major one being sold under the name of N-Serve. Nitrification inhibitors slow the conversion of ammonium-N to nitrate-N thus getting longer use out of a fertilizer application and possibly saving a subsequent application. There have been many reports on the benefits of these materials; there have been some reports that show no benefits. Because of the variables that affect the efficacy of these materials they should be tested on a particular piece of ground for a particular crop to see whether they will benefit that crop.

Nitrite Toxicity

Nitrite is an intermediate on the conversion of ammonium to nitrate. The microrganisms responsible for the conversion of nitrite to nitrate can be inhibited by high pH, high lime and high ammonium concentrations with the result that nitrite can accumulate in the soil to the point where it becomes toxic to plants. (The nitrification inhibitors described above inhibit the conversion of ammonium to nitrite, a separate process). Field reports of nitrite toxicity are uncommon, but it undoubtedly occurs and fieldmen should be aware of this possibility. Another advantage of the nitrification inhibitors is that they reduce, and probably eliminate, the possibility of nitrite toxicity.

General Reference

Bartholomew, W.V. and F.E. Clark (Editors): **Soil Nitrogen.** A monograph published by the Amer. Soc. of Agronomy, 677 S. Segoe, Rd., Madison Wis. 53711. 1965.

6. PHOSPHORUS (P)

Phosphorus (P) fertilization is necessary in order to get economic yields on many crops. Crops vary in their response to phosphorus; responses to phosphorus by tree and vine crops are virtually unheard of while phosphorus fertilization is a standard practice on many vegetable crops because it more than pays it's way.

Phosphorus is used mainly on row crops and here it's greatest beneficial effect to the plant is in the seedling stage and in the early stages of plant development. In most situations, phosphorus must be in the soil before or at the time of planting. Once the crop is well established, the chances for a phosphorus response are slim - side dress applications are rarely beneficial.

Phosphorus can promote earlier, more uniform maturity of crops, important in minimizing the chances of crop exposure to attack from pests, diseases and weather.

Deficiency Symptoms

There are no distinct, characteristic phosphorus deficiency symptoms for any crop. P deficient plant are generally more spindly and have a darker green color then plants with adequate P; crop maturity can be delayed. These symptoms are far from striking and as a result, phosphorus deficiency can easily go unnoticed.

Factors Affecting Phosphorus Response and Uptake

A number of factors have a strong influence on the chances of getting a crop response from phosphorus. The 6 major influencing factors are:

1. **Amount of phosphorus already in the soil**
   There is much less chance of a phosphorus response on a soil testing high in phosphorus than on one testing low.

2. **Soil moisture**
   Phosphorus moves more easily and is taken up more easily in moist soils than in drier soils. A good irrigation

can sometimes result in as much P uptake as an application of phosphorus.

3. **Soil temperature**
Phosphorus is more soluble and plants can take up phosphorus more easily as soil temperatures increase.

4. **Soil pH**
Phosphorus is most available to plants at a pH of 6.5 to 7.0. If pH gets too low, phosphorus can be tied up by the metallic elements (iron, aluminum, zinc) that are much more active in acid soil; if pH gets much above 7.0, P can be tied up by calcium and magnesium.

5. **Lime and gypsum in the soil**
Free lime and gypsum in soils will make phosphorus less available by tying it up as insoluble calcium phosphates. Heavy gypsum application (such as is often done for potatoes) can tie up phosphorus.

6. **Amount of zinc and other metallic elements in the soil**
Metallic elements tie up phosphorus. Zinc-phosphorus interactions have been well established; over application of one can create a deficiency of the other.

There is a wide range in reported crop responses to phosphorus - a crop will respond to phosphorus on one field, but not on another one which might be just across the road, even though soil types may be the same on both fields. Such response differences can usually be traced to one or more of the above 6 factors.

Written reports of crop yield increases from phosphorus fertilization should quantify each of the above 6 parameters to make the report meaningful, but this is rarely done. The fieldman should be leery about making too wide an application of the results of one test. Because so many variables can affect crop response to phosphorus, the only sure way to determine if a crop response is possible on a particular piece of ground is to apply phosphorus to that particular piece of ground.

Soil Tests for Phosphorus

Soil tests for P can give very good information on which to base a fertilizer recommendation but they are not definitive. For example, a soil reading of 15 ppm P might denote an excellent P level if the soil was warm at planting time, but the same level might be low if planting was done in the winter. A good deal of judgement must be used when evaluating soil P tests.

Many crops will respond to P in the starter fertilizer regardless of the soil levels of P mainly due to the fact that the seedling stage is responsive to P because the root system is not well enough developed to forage efficiently for P.

Placement of P

The sum of the knowledge on placement of phosphorus fertilizer was best stated by worker David Dibb: "If all of the P placement experiments in the world were laid end to end, they would all point in different directions."

Phosphorus does not move in the soil, but must be incorporated to come in contact with plant roots. Usually banding is superior to broadcasting-disking. This is especially true on soils that have a high fixing capacity for phosphorus (e.g., calcareous soils). Much of the variation in P placement results can be attributed to the P fixing capacity of soils. Banding is also more effective when soil temperatures are cool.

Optimum location of the fertilizer band will vary with the crop; it is desireable to have seedling roots intercept the band, thus placement would differ between a tap rooted crop and a lateral rooted crop. For example, a common band placement recommendation for tomatoes (a tap rooted crop) is 2 inches below the seed, while for potatoes it is 2 inches on either side. Using ammonium phosphate fertilizers or banding an ammonium form of N along with the P can significantly increase P uptake because of the acidifying effect that ammonium has on the soil when it is nitrified (converted to nitrate).

When soil P levels are very low (as shown by soil test) it will usually not be possible to band enough material to get a maximum response, and broadcasting most or all of the P fertilizer is recommended. Broadcasting P fertilizers then folding them into the beds is a widely used practice in California and is a cross between broadcasting and banding.

The fieldman should know the characteristics of the particular piece of ground he is working with to be able to make meaningful suggestions on P placement.

Improving Phosphorus Nutrition

Naturally, phosphorus fertilizers should be applied when a need can be demonstrated, however, if it is known that P responding crops are to be continually grown on a particular piece of ground it would pay to do everything possible to turn that piece of ground into a high P supplying soil. This does not necessarily mean adding P; an adjustment in soil pH can have more beneficial effect on P nutrition than the addition of phosphorus itself. Attempt to get the field into the 6.5 to 7.0 pH range through the use of ammendments.

On highly calcareous soils it may be impossible to reduce pH below 7.2 (e.g., it could take the equivalent of well over 100 tons of sulfuric acid to do so). Band application of acid or acidifying materials could be used on such soils; such applications have been shown to give a P response.

Manure, particularly poultry manure, can be a good supplemental source of phosphorus.

General Reference

Phosphorus for Agriculture, a 158 page manual published by the Potash/Phosphate Institute (1649 Tullie Circle, N.E., Atlanta, Ga. 30329). 1978.

Khasawneh, et al. (ed.) **The role of phosphorus in agriculture.** Amer. Soc. of Agron., Madison, Wis. 1980. (928 pp; $25).

43

7. POTASSIUM (K)

Agricultural crops require and use significant amounts of potassium; for example, a 30 ton crop of tomatoes or an 8 ton crop of hay each remove close to 500 lbs of K_2O from the soil.

Symptoms of potassium deficiency are characteristic for individual crops but often include yellowing and scorching of foliage. Although the evidence is not solid, potassium is felt to increase plant tolerance to disease, improve quality and shelf life of crops and improve red coloring of crops such as apples and grapes. Crop load is often directly associated with K deficiency, with prunes and tomatoes showing K deficiency only when heavily cropped.

The exact roles of potassium in plants are not well defined but K is believed to be involved in photosynthesis, carbohydrate and protein metabolism and water relations of plants.

There are 4 general classes of soil potassium:

1. **Mineral potassium** comprises 90 to 98% of a soil's total K supply. These minerals are very resistant to weathering and release negligible K to crops.

2. **Slowly available potassium** comprises from 1 to 10% of the total soil K. It is held tightly between clay minerals and is released only slowly to plants.

3. **Exchangeable potassium** comprises about 1% of the total K but is available to plants.

4. **Solution potassium** is available to plants but also susceptible to losses from leaching; it represents about 1/2% of the total soil K.

Potassium adsorption by crops is adversely affected by cool soil temperature, compacted or poorly aerated soil and ammonium-N nitrogen. Thus a crop growing on a cool, poorly aerated soil and heavily fertilized with ammonium sulfate would be a candidate for K deficiency.

Potassium-Water Relations

Crops can show deficient K levels in leaves and exhibit incipient symptoms of K deficiency due to drought rather than a shortage of soil K. K levels in plants rise after an irrigation. Keeping a crop well supplied with water can improve K nutrition better than application of K fertilizers.

Soil Analysis for K

Soil analysis is routinely run for K in most areas and many fertilizer recommendations are made on the basis of such analyses. Because of the different forms of K in the soil (see preceding) it is difficult to get a soil extraction that will measure solely the amount of K available to plants. The validity of soil K tests as an index to predict K needs of crops is currently coming under serious questioning. In Iowa, K is recommended for corn even on high K soils; in North Dakota and Montana, K responses have occurred on soils that have tested very high in K. South African scientists are currently re-evaluating soil K tests after concluding that "the conventional soil analyses do not present a reliable indication of the potassium supplying power of soils in all cases." California is now recommending a nitric acid soil extraction for K (to replace an ammonium acetate extraction) which looks promising. Because of the wide variation in soils it will be difficult to get a K test that is good for all soils.

Plant Analysis for K

Plant analysis for K can be a useful tool in diagnosing K problems, but it is not a definitive tool. As indicated previously, plants under stress for water often test low in K; applying potassium in such cases would be unwise.

A Purdue study suggested that the ratio of K between young and old tissue was a better indicator of K status than any absolute value. The study put it this way: "If the mature tissue contains a percentage of K equal or above that of the immature tissue, the K status of the plant is optimum. But if the mature tissue contains a lower percentage K than the immature tissue, the K status is less than optimum." If valid, this would be a good diagnostic tool.

A study on alfalfa indicated that ratios of K to N and P were a better indicator of K status than K levels alone.

Potassium Application

Soil application

Two materials are available for soil K application: Muriate of potash (61% K_2O) and sulfate of potash (51% K_2O).

Sulfate of potash (or potassium sulfate) is more expensive than muriate of potash (or potassium chloride) in spite of its lower K analysis. Some growers are leery of the chloride in muriate so they pay more for and use the sulfate form. In most cases, muriate is just as safe as sulfate when broadcast; band application should be avoided and the sulfate form should be used, if needed, on crops that are extremely sensitive to chloride such as strawberries and avocados.

If sulfur is also needed to correct a sulfur deficiency, then the sulfate of potash form is superior.

Soils vary in their potassium fixing capacity. There is generally very little movement of potassium (from the point of application) on heavy textured soils while movement can be significant on coarse textured soils. Calcium displaces potassium on soil exchange sites and the application of gypsum (calcium sulfate) in conjunction with potassium has been shown to give significant downward movement of potassium. Irrigation water that is high in calcium should also give significant downward movement of potassium. On a drip irrigation experiment, potassium sulfate applied directly under the dripper moved down 2½ feet on a soil with a relatively high K fixing capacity. For tree and vine crops, potash fertilizers applied in a concentrated band should move downward significantly on coarse textured soils, esp. if irrigation water is high in calcium. On heavy textured soils, deep placement may be necessary.

If high amounts of potassium are needed for row crops, most or all of the material should be broadcast (and disced in). Once soil levels of K have been built up over the years, the amount of K in a triple mix fertilizer should be sufficient to satisfy the needs of row crops with a high K requirement, such as potatoes. Because of its higher analysis (compared to sulfate of potash), muriate of potash is used in most triple mix fertilizers - fertilizer burn (from chloride) can occur when large amounts of such mixes are banded on row crops.

Foliar application
Because soil applications are more economical and provide a greater amount of K, foliar applications of K are not common. When foliar applications are used, potassium nitrate is the material used. Potassium nitrate sprays are frequently used on prune trees in the Sacramento valley of California in those years when very heavy crops induce a K deficiency; such sprays must be repeated at about 2 week intervals to have a continued beneficial effect.

Effects of Excess Potassium
Too much potassium is not directly toxic to plants but can have adverse effects. Over fertilization with potassium can induce a magnesium deficiency. On apples, too much potassium in relation to calcium will increase the severity of bitter pit. Caution should be used when applying potassium; it should be applied only where a need can be shown.

General Reference
Kilmer, V.J. et al, (ed.). **The role of potassium in agriculture.** Amer. Soc. of Agron, Madison Wisconsin. 1968. (Proceedings of a 1968 symposium on K published in book form).

8. SULFUR - Who needs it?

From the information on sulfur that has been disseminated in recent years, a fieldman could easily conclude that if the growers he worked with didn't already have sulfur deficiency problems they probably soon would.

Sulfur is needed by plants in about the same quantities as phosphorus. High protein crops such as alfalfa have a relatively high sulfur requirement because sulfur is a major constituent of protein. 10 tons of alfalfa contain about 50 lbs of sulfur. Other crops can remove from 10 to 80 lbs of sulfur per acre from the soil.

In spite of the known crop needs for sulfur in relatively large amounts, it is surprisingly difficult to get a good handle on this nutrient. Neither soil nor plant analysis values have been refined to the point where critical levels can be stated with confidence. Plants analysis for sulfate is felt by some to be superior to analysis for total sulfur but neither gives a precise index of S status. N:S ratios in plants can provide clues, with a high N:S ratio indicating a potential S deficiency problem.

The best and only definitive way to confirm a suspected sulfur deficiency is to see if a crop response can be obtained by the application of sulfur.

Those that are in the business of selling sulfur to agriculture continually hammer away at 2 points:

1. Increased use of high analysis, "purer" fertilizers has resulted in a great reduction in sulfur that was previously added to crops.

2. Increased pollution controls have resulted in a drastic reduction in the amount of sulfur contributed to the air (and subsequently to the soil) by industry.

The first point is valid, however it should also be brought out that sulfur used in pest and disease control programs can easily satisfy a crop's sulfur needs. For instance, vineyards on which sulfur is used to control mildew will never be sulfur deficient.

The second point is open to question as it does not square with the increased reports of acid rain (largely caused by SO_2) in many parts of the world. This apparent contradiction - increased pollution control + increased reports of acid rain - is due to the fact that much of the pollution control is directed at controlling pollution at and around the source. What industry has done is to build higher smoke stacks that spew SO_2 into the upper atmosphere. This SO_2 is carried many miles from the source; rain brings the SO_2 down in the form of sulfurous and sulfuric acid (acid rain). Pollution controls have affected the distribution of pollutants more than the actual amounts. If SO_2 is directed away from population centers, then that much more will fall on agricultural land.

How much sulfur does acid rain contribute to agriculture? Probably between 5 to 20 lbs per acre in affected areas - enough to satisfy crop needs. A good question is, where are the affected areas? It's not all that hard to monitor S in rain water and logging of S content of rain by pollution control agencies will likely become a standard practice before very long. Fieldmen can use such data when planning S fertilization programs.

Sulfur Contributions From Irrigation Water

When reading a report on an observed crop response from sulfur, the first thing the fieldman should look for in the report is the **sulfur content of the irrigation water.** One authority has calculated that "Responses to sulfur fertilization can be expected only where less than 5 lbs per acre of the element is supplied with the irrigation waters." A detailed study in Washington state suggests that a response to sulfur can be expected only if the irrigation water source contains less than 5 lbs of S per acre foot (a concentration of about 5.5 ppm sulfate). Most wells contain this amount of sulfur and many rivers do also. Relatively pure water from mountain streams (and snowmelt from the Sierras) can have a very low sulfur content. It is crops that are irrigated with **these** waters (or not irrigated at all) that are candidates for sulfur deficiency.

It is very surprising that in all the reports of sulfur deficiency, the influence of the sulfur content of the irrigation water is rarely mentioned. When one reads of thousands of acres responding to sulfur it often turns out that the crop is non-irrigated pasture. Sulfur responses on irrigated agricultural crops are not common; when they do occur, the water source pretty much has to be low in sulfur, but you usually wouldn't know this from reading about the response.

Correcting Sulfur Deficiencies

Gypsum (calcium sulfate) application is probably the best method of correcting a sulfur deficiency, esp. on irrigated crops. If a sulfur deficiency occurs on irrigated crops it will invariably be because the irrigation water is low in sulfate. Low sulfate irrigation water is usually low in total salts (and in calcium); infiltration problems can occur with such water, thus the gypsum can serve the dual purpose of adding sulfur and improving water penetration.

Elemental sulfur can be used to correct sulfur deficiency but since elemental sulfur is acidifying, soil pH control will be a concern on non-calcareous soils (most sulfur deficiencies occur on non-calcareous soils). Incorporating sulfur in an insect or disease control program is a good way to get sulfur to a crop. Responses from elemental sulfur will not be immediate as the biological conversion of sulfur to the usable sulfate form is a process that can take up to a year or more.

Using ammonium sulfate as the nitrogen source for a crop is a good method of adding sulfur. Since ammonium sulfate is an acidifying material, pH control will be important.

For row crops that require phosphorus, a sulfur-containing phosphorus fertilizer (e.g., 16-20-0) is a good choice.

Sulfur materials are not limited to the 4 materials mentioned above, but these 4 are the most widely used and most practical materials for correcting sulfur deficiency.

General Reference

The Sulphur Institute (1725 K St., N.W. Washington, D.C. 20006) started an annual publication, **Sulphur in Agriculture** in 1977. This publication gives annual summaries and writeups of the significant work being done with sulfur in agriculture.

9. CALCIUM - Emerging Superstar

Calcium will be the most studied nutrient for the balance of this century, if it is not already, and over the next 100 years more calcium will be applied to agricultural crops than any other nutrient.

A good deal of the credit for the focus on calcium goes to USDA worker C.B. Shear who, in a 1975 paper, condensed much of the significant information on calcium into a little more than 3 pages followed by over a hundred references. Shear's short tract is truly a treasure trove of information and should open many eyes to the magnitude of potential calcium related disorders - it can already be termed a classic. Shear also spearheaded an international symposium on calcium in 1977 that has provided an additional forward thrust to calcium related research.

It is understandable why calcium has not been given more consideration than it has. Most textbook discussions of calcium deficiency draw from nutrient culture studies in which plants are grown without calcium. Under these conditions cell wall structure deteriorates, providing classic pictures of drooping plants - such pictures are permanently etched in the minds of plant nutrition students.

These textbook symptoms of calcium deficiency are simply not seen in the field. Couple this with the fact that calcium is the 5th most abundant element in the earth's crust and it is little wonder that calcium nutrition is the furthest thing from the minds of fieldmen when dealing with day to day problems in agriculture. When working on a crop disorder occurring on calcareous soils* it is little wonder that calcium nutrition is not even considered.

The paramount role currently assigned to calcium is maintaining the structure of membranes. Calcium counteracts the potential harmful effects of other divalent cations and also of sodium and

---

*calcareous soils are defined as those containing calcium carbonate (also referred to as lime); calcareous soils are extensive in arid areas.

boron. One researcher put it this way: "The greater a demand is made on the selective machinery of the plasmalemma, through the presence in the medium of potentially damaging ions at high concentrations, the more crucial seems to be the role of calcium in maintaining the integrity of the membrane." Toxicities of other elements, including aluminum, are sometimes associated with calcium related disorders and such toxicities should be considered when attacking potential calcium related disorders.

In tackling calcium related problems, agriculturists should discard traditional thinking on nutrient deficiencies. One of the first steps towards understanding calcium nutrition is separating calcium "deficiency" from "calcium stress". A "deficiency" of calcium is a condition that is corrected only by the addition of calcium. "Calcium stress" is caused by a temporary localized inadequacy of calcium; the entire plant is usually not affected with the stress often confined to the fruiting part or growing point. Addition of calcium to the soil does not always correct calcium stress since the temporary, localized shortage of calcium may be beyond the plant's adsorption and mobilization powers to correct. Environmental factors can have more influence on calcium stress disorders than actual amounts of soil available calcium.

Foliar sprays of calcium (usually calcium nitrate or calcium chloride) are often more effective than soil applications but even foliar sprays do not always correct calcium related disorders. Soft apples and bitter pit of apples, both calcium related disorders, are better corrected by post harvest treatment of apples than by soil and/or foliar applications of calcium.

A study of calcium related disorders shows that many are associated with moisture relations in crops. There are hundreds of research papers dealing with blossom end rot in tomatoes and peppers - authors of some of these papers present convincing evidence that calcium nutrition is the causal factor and other authors present equally convincing data that moisture stress is the sole cause; probably both schools of thought are correct. Because of the precipitation of calcium compounds as the soil dries, moisture stress may automatically mean calcium stress. Calcium: sodium ratios of soil solution are also involved.

In an experiment on a highly calcareous clay soil with an ESP of 7.3, frequent (daily or twice daily) sprinkler irrigations gave a 4% incidence of blossom end rot vs. a 30% incidence on the control. The differences are likely related to Ca:Na ratios in the soil solution.

Epstein has stated that "although all growing points are sensitive to calcium deficiency, those of the roots are affected most severely." Calcium has been shown to improve the rooting of cuttings, primarily by improving (increasing) the calcium:sodium ratio. Research has shown that calcium **deficiencies** can occur on **calcareous** soils that have a high percentage of exchangeable sodium and other studies indicate that the ratio of calcium to other soluble salts is more important than the absolute value of calcium itself.

Improving the Ca:Na ratio (or reducing the SAR) in irrigation water and/or in soils has been shown to improve soil structure and water penetration and thus benefit plant growth. It is likely that improvement of the Ca:Na ratio beyond that necessary for soil structure improvement will benefit some plant species. Since calcium compounds are the first to precipitate out of the soil solution as the soil dries, esp. in calcareous soils, the Ca:Na ratio of the soil solution is not constant (esp. in calcareous soils) but is significantly higher in wet soil than in drier soil. The constantly recurring association of calcium related problems to moisture stress can be explained in part by Ca:Na ratios in the soil solution.

The 1946 study that showed that calcium **deficiency** occurred on calcareous soils (that have a high sodium %) answers those that say "impossible" when it is suggested that physiological disorders on a calcareous soil may be due to calcium stress. It is of interest that the two calcareous soils on which calcium deficiencies were found in this study were from Oregon and Washington where calcium disorders on apples are widespread. If calcium deficiency can occur on calcareous soils with a high sodium % it is certainly probable that calcium stress disorders can occur on calcareous soils that have a moderate sodium content, especially in the lower ranges of soil moisture.

Although calcium analysis of affected plant parts can sometimes provide useful information, plant analysis for (total) calcium is generally of very limited value in diagnosing calcium related problems. Plants accumulate large amounts of calcium as relatively insoluble, biologically inert compounds, particularily calcium oxlate*. The calcium in these compounds is not readily translocated and therefore not available to the areas of need of the plant. As leaves get older, the amounts of these biologically inert compounds increase (calcium levels in leaves are used as an index of leaf age in the interpretation of citrus leaf analysis). Plant analysis for calcium has been of little value in diagnosing calcium related disorders because calcium analysis values that include biologically inert compounds are not meaningful. It is possible that calcium is required in only micronutrient quantities during certain critical stages of plant growth.

With their favorite tool, plant analysis for the target element, taken from them, many fieldmen do not pursue potential calcium related problems. Such problems are classified as "physiological disorders" an imprecise but not incorrect term. "Physiological disorder" has become an accepted euphenism for "we don't know the answer."

Why are more and more calcium related disorders being reported? An historical look at the origin of the affected crops provides some clues. Crops on which calcium related disorders are most common originated on soils that were high in calcium, many of them calcareous. These plants could be termed calciphilic, or calcium loving species. Grow these crops under the same conditions as they originated and there should be no calcium related disorders. However, grow the same crops under intensive irrigation with a low calcium irrigation water (e.g., snowmelt from the Sierras) and you will gradually deplete calcium levels in the soil. Plant species that have evolved over the eons on low calcium soils or acid soils do not exhibit calcium related disorders because they have learned to survive with a minimal amount of soil calcium, i.e., they are not calcium stress susceptible.

---

*plants produce oxalic acid during periods of high activity. Accumulation of high amounts of oxalic acid can cause toxicity. Calcium is probably needed to neutralize oxalic acid by forming calcium oxalate.

Soils that are low in calcium are often low in boron because the same factors that caused low calcium cause low boron. Calcium stress disorders often occur along with and are interrelated with boron deficiency disorders; both must be corrected to effect a cure to a particular problem.

Today (1980) there are a number of "mystery" disorders showing up on California crops that have not been seen before and that cannot be traced to any known pathogen. These disorders are more common on the east side of the great central valley and include:

> Pox on nectarines
> Die back on walnuts
> Shedding (or shattering) of walnut flowers
> Shattering (or shedding) of grape flowers
> Fruit drop on a number of crops.

These disorders are currently labeled "physiological disorders" and are causing considerable head scratching among researchers. Perhaps calcium will be implicated in these disorders at a future date. (Boron may also be involved - see Chapter 13).

Calcium nutrition still represents the largest unexplored area of plant nutrition - this in spite of the fact that there is already a wealth of information on calcium. Much of this information, however, is fragmentary and contradictory and inevitably leads to frustration on the part of anyone trying to obtain a solid grip on calcium nutrition. This frustration was expressed in the following way by an extension specialist from Oregon: "The list of what we don't know about the factors influencing calcium distribution is really much longer than I have indicated. Our ignorance is so profound and the literature so full of poor research and un-warranted conclusions that I would be tempted to advise young pomologists and physiologists to discard all their reprints of research papers and start over from scratch."

Many calcium related problems rest in the "no man's land" between the disciplines of plant science and soil science and an interdisciplinary approach is needed to solve these problems.

Significant progress in calcium nutrition will only be made after more fieldmen ask the question, "is it possible that calcium nutrition is involved?"

General References

1. Shear, C.B., **Calcium-related disorders of fruits and vegetables.** HortScience 10:361-365. 1975.

2. Shear, C.B. (editor), **International symposium on calcium nutrition of economic crops.** Communications in Soil Science and Plant Analysis, 10 (1 & 2): 1-502 (special issue devoted entirely to the calcium symposium and comprising 37 papers). 1979.

10. MAGNESIUM - The Minor Major

Alot of literature exists on magnesium deficiency in agricultural crops. When this literature is reviewed, it is found that most magnesium deficiencies occur on acid, sandy soils in areas of moderate to high rainfall. There are many reports of magnesium deficiency from the southeastern U.S. but comparatively few from the western U.S., although magnesium deficiency is a possibility on alkali soil. Magnesium is abundant in most soils and occurs in many minerals which, when weathered, continually contribute to the soil's Mg supply. Irrigation waters also can contribute significant magnesium to soils.

Of the cases of magnesium deficiency reported in the western U.S., most have been caused by heavy applications of potassium and/or manure (which is high in potassium). It has been well established that potassium reduces Mg uptake. In some cases Mg deficiency has been corrected simply by ceasing potassium fertilization.

Fieldmen in the western U.S. should be aware that Mg deficiency can exist but should also be aware that it is highly unlikely. Plant analysis for Mg can be useful in diagnosing a suspected Mg deficiency.

If magnesium is needed, it can be applied directly to the soil or by foliar spray. For soil application either dolomitic lime or magnesium sulfate can be used; dolomite lime is cheaper but is only effective on acid soils and there is evidence that on some acid soils it is not effective.

A grower that knows he will eventually have to apply lime to his soil for pH control would be well off to apply dolomitic lime with the idea that it can't hurt and it might help to maintain Mg levels.

Magnesium toxicity caused by a high Mg:Ca ratio can be a problem on soils derived from serpentine parent material. Adding calcium to the soil is recommended in such cases.

11. IRON

Unlike most deficiences, iron deficiencies are usually not related to a low amount of the element in the soil but are brought on by one or more other factors. Some of the factors that can cause or contribute to iron deficiency are:

Lime in soil (lime induced chlorosis)
Bicarbonate in soil and/or water
Cool soils
Wet soils
High phosphorus content of soils
High zinc (and Mn) levels in soil at low pH

In many instances, iron deficiency could be called by other names, such as lime induced chlorosis, bicarbonate chlorosis, and cool, wet soil chlorosis. It would still be iron deficiency, but the different names indicate that different methods can be used to attack the problem.

Many crops will show symptoms of iron deficiency (interveinal chlorisis of foliage) in the spring of the year when soils are cool and wet and the deficiency will correct itself as the weather becomes warmer. Mild iron deficiency during a cool spring has resulted in the sale of many pounds of nitrogen fertilizer in the past and will continue to do so in the future. A mild case of iron deficiency can be mistaken for nitrogen deficiency; the grower applies nitrogen fertilizer, the weather warms up, the plants turn green and the grower and the fertilizer salesman congratulate themselves for a job well done.

For crops known to be susceptible to iron deficiency, water management should be watched closely, especially during cool weather such as occurs in the spring. Don't apply water unless needed.

I observed a zinc induced iron deficiency on almonds a few years back that was aggravated by low soil pH. The soil was a loamy sand, the grower had applied zinc religiously 2 or more times a year and soil pH was in the 4.5 range. Zinc levels in the soil had built up over the years and the zinc was very soluble at the low pH. Although the tree symptoms were typical of iron deficiency (metallic nutrients compete for entry into the plant) it would not be improper to call them zinc toxicity symptoms. Lime application and cutting back on zinc corrected the problem.

Application of acidifying ammendments to the water and/or soil is sometimes helpful for lime and bicarbonate chlorosis.

Gypsum application can also be helpful in relieving iron chlorosis. Gypsum alleviates water logged soils and also reacts with bicarbonate and phosphorus to reduce their levels.

Diagnosing iron deficiency is best done visually - trust the eyes. Plant analysis for iron is of little help and can give misleading information. High phosphorus levels in plant tissue are often associated with iron deficiency; high potassium levels are sometimes associated with the deficiency.

Application of iron materials, either soil or foliar, should only be tried if all other methods of correction fail. Foliar sprays are the most common chemical method of correcting iron deficiency. Soil applications of chelated materials can provide correction in some cases, but are expensive. Materials other than chelates are rarely used for soil application because they are readily fixed in an unavailable form in most soils.

12. ZINC - The Major Minor

It is the opinion of many that after nitrogen, zinc is the most limiting nutrient toward achieving maximum crop yields. More and more zinc deficiencies are being reported - many from areas where zinc deficiency has not previously been reported. New, higher yielding crop varieties and improved fertilization practices of other nutrients, particularily nitrogen, are the major reasons for the increased reports of zinc deficiency; a greater awareness of zinc deficiency on the part of farmers and fieldmen and gradual depletion of soil zinc over the years are other reasons.

On many crops, zinc deficiency is the rule rather than the exception. For example, in the southern San Joaquin valley of California, I have never seen a young, vigorously growing almond orchard that did **not** show zinc deficiency symptoms.

Crops vary in their sensitivity to zinc deficiency as shown by the following classification:

ZINC DEFICIENCY CLASSIFICATION

Very Sensitive	Somewhat Sensitive	Somewhat Tolerant
tree and vine crops, beans, onions, corn, garlic	cotton, sorghum, tomato	grain crops, lettuce, potato, sugar beets, carrots, safflower, alfalfa

The above classification will undergo modification as more knowledge is gained on zinc nutrition.

Factors Influencing Zinc Deficiency

A number of factors affect the probability of a particular crop developing zinc deficiency. Some of the more important factors are:

1. **Soil pH**
   As soil pH rises above 6.0, zinc availability to plants decreases.

2. **Lime content of soil**
   Lime ties up zinc. Magnesium carbonate is more detrimental than calcium carbonate in this respect.

3. **Land leveling**
   Zinc content of soils decreases with soil depth. Cut areas of soil are very susceptible to zinc deficiency.

4. **High amounts of phosphorus**
   Phosphorus ties up zinc. Heavy phosphorus fertilization can create a zinc deficiency.

5. **Organic matter and manure**
   Zinc is less available in soils with a high organic matter content. Heavy manure applications can create a zinc deficiency - chicken manure is particularily bad in this regard. Historically, some of the first reports of zinc deficiency were from old corral sites.

6. **High levels of other metallic elements**
   Metallic elements compete with each other for entry into the plant. High levels of the other metallic nutrients (copper, iron, manganese) can induce zinc deficiency; a heavy application of iron material can induce a zinc deficiency (and vice versa).

7. **Cool, wet soils**
   Like phosphorus, zinc is less available to plants at cool soil temperatures. Row crops can be zinc deficient early in the growing season, then grow out of it as the weather warms up.

8. **Sunlight intensity**
   Plants are more susceptible to zinc deficiency in bright sunlight. This relationship is believed to be associated with auxin activity.

9. **Crop rotation**
   The previous crop influences the availability of zinc to the present crop. Zinc deficiencies on cotton are common following sugar beets and rare following alfalfa.

Variations in the above factors account for differences in observed crop responses to zinc.

Diagnostic Aids

One or more of three methods given below can be used to diagnose zinc deficiency:

1. **Soil Analysis**

   Soil test levels of zinc can give excellent information on the probability of a particulr crop getting zinc deficiency. Soil tests will not precisely predict a crop response (or lack of response) but they are recommended pre-plant for row and field crops. Because soil temperatures influence zinc uptake, an interpretation of soil zinc might be valid at one temperature, but not at another; e.g., a soil zinc level that may be optimum for a warm soil may be too low for a cold soil.

2. **Symptoms**

   Zinc deficient crops usually, but not always, exhibit characteristic deficiency symptoms. The most common symptoms are interveinal chlorosis or mottling of leaves, reduced terminal growth with small leaves and short internodes (rosetting) and in severe cases, leaf necrosis and growing point dieback. Plants in the seedling stage can be suffering from zinc deficiency but show no other symptoms than a general stunting; when a whole field is generally stunted it is difficult to diagnose a deficiency. Poor early season growth of row crops due to zinc deficiency is often attributed to other causes. Plants can grow out of early season zinc deficiency as the soil warms up, but overall crop maturity will be delayed due to the poor early season start.

3. **Plant Analysis**

   Plant analysis (usually leaf analysis) can provide supportive information when diagnosing zinc deficiency. High zinc levels (over 30 ppm) or very low zinc levels (below 10 pm) can confirm field observations on zinc status. Unfortunately most zinc levels of plant tissue fall in the gray area between deficient and sufficient (between 10 to 25 ppm) and no conclusions can be drawn. I have seen orchards that I have known were zinc deficient, that showed zinc deficiency symptoms and that responded to zinc application yet the leaf levels of zinc were well above 20 ppm.

Although the above 3 methods are useful in diagnosing zinc deficiencies, the only definitive method is to see if a crop response can be obtained by the addition of zinc.

Zinc Application

Zinc may be applied to the soil (the usual method for row crops) or as a foliar spray (the usual method for tree and vine crops). A variety of zinc materials are available. Material should be selected on the basis of performance and economy. Soil and foliar applications are discussed in the following:

Soil application of zinc

Application of approximately 50 lbs of 36% dry zinc sulfate or 17 gallons of 10% liquid zinc sulfate is a common method of applying zinc to row crop soils in California. One soil application can last up to 5 years or more. Liquid zinc sulfate is probably superior to the dry since widely spaced dry zinc particles do not contact a maximum number of roots.

There are other materials available for soil applicaton - chelates, organic complexes, zinc ammonium nitrate; with the exception of zinc sulfide, all are effective, however zinc sulfate has proven to be the most economical. All inorganic forms of zinc must be incorporated to be effective.

Zinc chelates are expensive, but can be both effective and economical if banded with other fertilizers; banded with phosphorus, chelates offset the depressing effect of phosphorus on zinc uptake. Zinc chelates applied in a small band around young trees can be used to supply zinc to small trees but this practice is not economical for large trees. Zinc chelate is being applied through drippers, apparently satisfactorily, although there is not yet enough solid evidence on this.

Although not usually tried, soil application of zinc sulfate may have merit for mature trees. Mature trees can have many roots in the top foot of soil that can pick up zinc. In a test on pecans on a calcareous soil, zinc sulfate, broadcast and incorporated to a depth of 6″ was picked up by the trees; application of sulfur in conjunction with the zinc enhanced zinc uptake. In a walnut orchard, zinc applied in a trench around a tree and irrigated in was found to move 2 to 4 feet downward. The chemical makeup of the irrigation water undoubtedly influences the downward movement of zinc; more downward movement can be expected with a high calcium well water than with a low salt river water.

Foliar application of zinc
Foliar sprays of zinc materials, applied alone or in combination with insecticides, are widely used for both zinc deficiency correction and for zinc nutrition maintenance. A wide variety of zinc materials are used in foliar sprays including zinc oxide, basic zinc sulfate, zinc sulfate + lime (as a safening agent), organic zinc compounds including chelates and safened zinc proprietary materials.

Many of these materials have inherent disadvantages - zinc oxide is of low solubility and may not be as effective as other materials, basic zinc sulfate is not compatible with many insecticides and is relatively expensive, zinc sulfate + lime is not compatible with many insecticides because of the lime, organic zinc compounds and proprietary zinc materials are relatively expensive. Although relatively costly, the chelated zinc materials have given good results.

Straight zinc sulfate foliar sprays have not been widely used in the past because of the possibility of phytotoxicity (leaf and fruit burn). University of California recommendations for nutrient foliar sprays on citrus as recently as 1966 were to add lime or soda ash to all zinc sulfate sprays to safen the zinc. It was later found that a significant safening effect could be obtained without lime simply by cutting back on the amount of zinc sulfate used. It was found that low concentrations of straight zinc sulfate (36%) had the same active zinc concentration in solution as higher rates of zinc sulfate + lime. All the lime was doing was taking some of the zinc out of solution; the same thing could be accomplished by not using as much zinc to start with. Although it is not discussed widely today, the benefit of hindsight shows that the old lime-zinc recommendations were silly, and a waste of both lime and zinc.

There is currently much interest in the use of low concentration zinc sulfate sprays. Such sprays can have the advantage of being both effective and economical. These sprays offer growers two sizeable advantages:

1. lower per acre material cost

2. compatibility with most insecticides; straight zinc sulfate can **increase** the effectiveness of many insecticides by lowering the pH of the spray solution (this effect will also reduce or eliminate the need for buffers in the spray solution with many water sources).

So far, tests with these sprays are very promising. Current recommendations are 1/4 lb of 36% zinc sulfate per 100 gals for stone fruits to 1 lb/100 gals for citrus, grapes, walnuts and row crops. Zinc sulfate sprays should not be applied on crops such as lettuce where foliage spotting is detrimental.

If a grower is aware of the potential hazards of leaf burn and fruit spotting, he should be able to use zinc sulfate sprays effectively. Where spotting of fruit is a major concern, the lower rates should be used or test plots only should be tried. Foliage spotting may be unsightly but plants will grow out of it.

The following points should be kept in mind when using zinc sulfate sprays:

1. Material should be accurately mixed to avoid overdosage.

2. Material should be thoroughly dissolved in spray tank. Use of zinc sulfate solution usually solves this problem.

3. Urea increase the absorbtion of zinc by leaves. When urea is included in the spray mix, use the lower zinc rates.

4. Zinc sulfate is highly corrosive. Flush equipment thoroughly after use.

5. Spraying in cool or inclement weather or just before a rain increases the burn hazard.

Other inorganic zinc sources at low rates are also being marketed for use as zinc foliar sprays. A zinc nitrate material mixed with a nitrogen fertilizer (sold as N-Z-N) has shown to be superior to zinc sulfate on pecans but further testing is needed with this material; as with zinc sulfate, this material can also cause phytotoxicity.

For any foliar zinc application it is best if the spray wets the foliage thoroughly. Concentrate sprays (20 gals/acre) have been succesfully used on citrus with no burn at the rate of 1.7 gals of 10% zinc sulfate or 5 lbs of 36% zinc sulfate per 20 gals. Aerial spraying has been effective on many crops.

Because of the widespread need for zinc on many crops, growers and fieldmen should attempt to work zinc in with any pesticide sprays that go on during the growing season. If leery about phyto from zinc sulfate sprays, then zinc chelate sprays should be used.

Zinc Toxicity

High soil concentrations of zinc can be toxic to plants. Zinc toxicity is more of a concern on acid soils because of the higher solubility of zinc on acid soils. Soil buildup of zinc from foliar sprays over the years can be significant. Most almond growers put on a fall zinc spray containing roughly 30 lbs of 36% zinc sulfate. Put this much zinc on every year for a period of 20 years and you are adding a considerable amount of zinc to the soil. On an acid, sandy soil, zinc toxicity on an older almond orchard should not come as a surprise. (Almond growers should pay closer attention to the actual need for zinc rather than apply it routinely as many do).

With the current heavy emphasis on zinc in agriculture care should be taken not to get over zealous with zinc applications. Determine if a need exists before applying zinc.

The antidote for zinc toxicity is the application of lime.

General References

1. Seatz, L.F. and J.J. Jurinak. **Zinc and Soil Fertility.** USDA Yearbook of Agriculture, 1957. p 115-121.

2. Viets, Frank G., Jr. **Deficiency of field and vegetable crops in the west.** USDA Leaflet No. 495. 1961 (revised 1967).

3. **Diagnosis and treatment of zinc deficiency in crops.** American Zinc Institute, Inc., 292 Madison Ave., New York, N.Y.

4. Thorne, W. **Zinc deficiency and its control.** Advances in Agronomy, 9:31-65. 1957.

13. BORON

As most fieldmen are aware, boron is needed by plants in only very minute amounts and the range betwen boron deficiency and boron excess is not great. Boron deficiency and boron toxicity are discussed separately below.

Boron Deficiency

As more and more irrigation project water is used (as opposed to well water) we will be seeing more and more boron deficiency. This is because most project water comes from sources that are low in boron (e.g., snowmelt from the Sierras). Continued use of this water will gradually deplete boron in the soil - this depletion will be more rapid on sandy soils. Boron deficiency could well be involved in some of the "mystery" maladies listed in the chapter on calcium.

When attempting to diagnose a suspected boron deficiency, the first thing the fieldman should do is look at the boron content of the water source. If the B content of the water is 0.3 ppm or greater, you can pretty well eliminate the possibility of boron deficiency (with an exception to be given later).

Boron deficiency affects mainly the growing points of plants or the fruiting parts of plants. If terminal dieback is seen, if there is irregular spotting, discolorations or lesions on fruit, suspect boron deficiency.

An excellent example of the limitations of plant and soil analysis in diagnosing boron deficiency comes from the Pacific northwest. Prune and pear growers in this area have found that fall sprays of boron significantly increase the set of fruit the following spring. This, in spite of the fact that leaves show adequate levels of boron and that in some cases soil levels of boron are "dangerously high". Without the fall boron sprays, the trees do not exhibit classic boron deficiency symptoms - there was simply a higher than normal flower drop in the spring that could easily have passed unnoticed or have been attributed to other causes - a good example of "hidden hunger".

Correction of boron deficiency

An idea of the very small amount of boron needed to corect boron deficiency can be appreciated when it is realized that 1 ppm boron in irrigation water is a toxic level for many crops. 1 ppm boron in irrigation water supplies 8 to 10 lbs of boron annually, thus only 8 to 10 lbs of actual boron per acre (for 1 year only) should be ample to corect a deficiency. 10 lbs of boron per acre is 1 ounce per 272 square feet.

Boron materials are available for both soil and foliar application. The actual boron content of any material should be checked closely to make sure that an excessive amount is not applied.

Boron Toxicity

Boron toxicity is far easier to diagnose than boron deficiency. Areas of potential boron toxicity have usually been fairly well defined; if not, soil analysis can define them. Soil and water analysis can provide excellent supportive information (plant analysis is also useful, but not always definitive) when diagnosing boron toxicity.

There is no antidote for boron toxicity. The only recommendation is plenty of water and leaching. Sulfuric acid has been found to greatly increase the rate of boron leaching from high boron soils.

Soils differ in their capacity to adsorb boron from or release boron to the soil solution. Soils will therefore differ in the amount of time and leaching that will be necessary to reduce boron to safe levels.

Calcium Effects On Boron

A 1944 study showed the following effect of calcium and potassium on boron deficiency and toxicity:

Nutrient	Effect on boron deficiency	Effect on boron toxicity
Calcium	Increases severity	Decreases severity
Potassium	Increases severity	Increases severity

Thus, in a situation where both calcium stress and boron deficiency co-exist (a not unlikely set of circumstances) the addition of calcium alone could well increase the severity of crop symptoms caused by boron deficiency.

General Reference
Gupta, U.C. **Boron nutrition of crops:** In: Advances in Agronomy, Vol. 31, 1979. Academic Press, 111 5th Ave., N.Y., N.Y. 10003.

14. THE MINOR MINORS
(Manganese, Copper, Molybdenum, Chlorine)

Manganese

Manganese (Mn) deficiency is not nearly as common as zinc deficiency but can be a problem on calcareous soils and on soils with an alkaline pH. Since the metallic nutrients (zinc, iron, copper, manganese) compete with each other, an excess of one can induce a deficiency of another. Zinc and iron induced manganese deficiencies are certainly possible when large amounts of either zinc or iron are applied. In case of deficiency, manganese can be applied either directly to the soil or by foliar spray.

Manganese toxicity is more common than manganese deficiency. Mn toxicity is often one of the first disorders to appear when soils become too acid. Manganese levels in leaves can be used to monitor soil pH; as soil pH drops, Mn levels in leaves will increase. Poorly drained soils are also candidates for Mn toxicity.

Copper

Copper deficiencies are not common but can occur on peat soils and on very sandy soils (and occasionally on other type soils). Copper deficiencies are usually associated with specific soil types; unless copper deficiency has been previously reported on the particular soil type in question it is very unlikely that copper deficiency will be a problem on that soil.

Plant analysis is not a good method of delineating copper deficiency because of the very low levels of copper that are found even in healthy plant tissue. The only definitive method of diagnosing a copper deficiency is to see if a response can be obtained by the addition of copper.

Copper toxicity can be a problem in some cases. Extensive spraying with copper fungicides can cause a soil buildup of copper over the years that can be toxic, esp. if the soil is acid.

Molybdenum

Molybdenum (Mo) is needed in less quantity than any other plant nutrient, yet field Mo deficiencies have been reported. Mo deficiencies almost invariably occur on acid soils and usually liming of the soil will correct the problem just as easily as applying molybdenum. Mo deficiency is highly unlikely on alkaline and/or calcareous soils but is possible on soils derived from serpentine parent material.

Molybdenum toxicity of crops does not occur, but high molybdenum levels in plants used as livestock feed can create a livestock poisoning known as molybdenosis. Molybdenosis is especially acute at low copper concentrations and is felt by some to be molybdenum induced copper deficiency. It is usually corrected by supplementing feed rations with copper.

Chlorine

Although chlorine is an essential element for plants, no cases of field deficiencies are known to the author. A possibility of chlorine deficiency has been suggested for sugar beets grown in Washington. When diagnosing a field problem, the fieldman can pretty well forget about chlorine deficiency being involved.

PART III:

OTHER
CONSIDERATIONS

15. POTENTIALLY TOXIC NON-ESSENTIAL ELEMENTS (Al, Cd, Cr, Fl, Li, Ni, Se, Si)

Aluminum

After oxygen and silicon, aluminum is the most abundant element in the earth's crust (soils range from 4 to 14% aluminum oxide). Aluminum is adsorbed by plants just like any other element and, like the metallic nutrients, its concentration in the soil solution increases as the soil becomes acid (as the pH drops). Many of the symptoms seen on plants growing on excessively acid soil are symptoms of aluminum toxicity. Aluminum is also associated with calcium deficiency. Unlike many of the essential plant nutrients, aluminum is rarely run on plant tissue on a routine basis. Aluminum should be run more often since it affects the balance of metallic nutrients and affects calcium nutrition. A history of routine aluminum analysis on plant tissue for a particular crop could be enlightening. Raising the soil pH is the best method of countering aluminum toxicity; application of phosphorus (which ties up aluminum) can also help.

Cadmium

Cadmium is a metallic element that is used in industry; sewage sludge can contain toxic levels of cadmium. As more sewage is used in agriculture, this potential problem will be monitored more closely.

Chromium

Chromium toxicity can occur on soils derived from serpentine parent material.

Fluoride

Fluoride is an impurity that can occur in significant quantities in some lots of superphosphate fertilizer. Fluoride toxicity has occurred in greenhouse culture and has been found in toxic levels in planting mixtures and in water.

Lithium

Lithium can occur in toxic amounts in well waters in the southwestern U.S.

Nickel

Like chromium, nickel can be found in toxic amounts on soils derived from serpentine parent material.

Selenium

Selenium can be toxic to plants at high concentrations. There are soils that are naturally high in selenium (seleniferous soils). Interestingly, although not essential for plants, selenium deficiency can cause problems in livestock; these problems can occur when a feed source that is low in selenium is used.

Silicon

Although not considered toxic to plants, silicon is mentioned here because of the vast quantities found in soils. Next to oxygen, silicon is the most abundant element in the earth's crust; soils range from 60 to over 90% silicon oxide. Silicon is not an inert material and can effect the nutritional balance of plants. Silicon has been shown to improve rice yields and may one day be shown to be essential for some plant species.

16. SOIL ANALYSIS

Used wisely, soil analysis can be very valuable in determining optimum fertilizer programs. Soil analysis has its best application as a pre-plant guide for row and field crops, but can also be useful in orchard and vineyard crops (see later).

Soil tests are available for most plant nutrients; test results for some nutrients are more reliable than others. The most widely used soil tests are for P, K and Zn; these tests can provide excellent guidelines for planning a fertilizer program (soil K tests have limitations - see chapter on K).

Soil tests for nitrogen should be interpreted with caution. There are many forms of N in the soil, some of them of very low availability to plants; biological activity causes changes in the forms of N in the soil. Nitrate-N is readily available to plants and a soil nitrate test can give good pre-plant information if soil is sampled shortly before planting. For years, the University of California refrained from recommending soil nitrogen tests because of the interpretation hazards but U.C. has recently come out with the following tentative guidelines for soil nitrate:

N RESPONSE PROBABILITY AS RELATED TO SOIL NITRATE LEVELS

| | — ppm Nitrate Level in Soil — | |
Probability of N Response	Cool season Crops	Warm season crops
Response likely	0 - 15 ppm	0 - 10 ppm
Response unlikely	over 25 ppm	over 15 ppm

The above table will undergo modifications and refinements as additional information becomes available. It should be remembered that the first irrigation after sampling will change soil nitrate values significantly and a follow-up sample could be helpful. Also, high soil nitrate is often associated with high salinity (EC) levels. With high soil salinity, nitrate is less available to plants; therefore a high soil nitrate reading at a low soil salinity indicates a better soil nitrogen status than a high soil nitrate level at high salinity.

Soil tests are also useful in pH control, salinity-alkalinity diagnosis and detection of toxic elements, especially boron. Knowledge of the lime content of soil is necessary in selecting proper soil ammendments, if needed. Some fieldmen, myself included, carry around a dropper bottle of dilute hydrochloric acid for a quick field evaluation of lime content - a drop or 2 of acid will cause soil to fizz if lime is present in significant quantity (muriatic acid, 1:10 dilution can be used as can lemon juice or vinegar).

Soil Sampling

The most accurate soil analysis in the world is no better than the sampling method used to take the sample. Take care to get a good, representative sample. For pre-plant analysis on row crops, 20 to 30 1" diameter cores taken to a depth of 8 to 12" should be composited to make about a 1 quart sample for an area. When sampling to diagnose problems, take the sample from the area where the problem is occurring. For example, if a germination problem exists, sample soil from around and just below the seed - this might mean the 2 to 4" level. If a water penetration problem exists, the top inch, or possibly the top millimeter of soil should be sampled.

When sampling for salinity-alkalinity, remember that hot spots (or alkali spots) can occur; take care to sample such areas separately. Better yet, get an aerial photo of the field during the growing season. Such a photo can aid in sampling and can eliminate the need for much sampling and analysis since it can delineate the hot spots needing attention.

Analysis Methods

The wide variety of soil analysis methods used can cause bewilderment when it comes to interpreting results. The best bet is to stick with the analysis methods recommended in your area - consult your extension service for this information. Insist that the lab use the recommended methods, or else take the sample to another lab.

There is some sentiment towards standardizing soil test methods throughout the U.S. There are difficulties involved in this, but more uniformity than presently exists would be desireable.

Soil Analysis for Orchards and Vineyards

The University of California's opinion on the value of soil analysis for orchards (and vineyards) has been, and is, that it is useful for pH, salinity-alkalinity, and toxicity (Na, Cl, B) determinations but is of little value in assessing nutrient status because of the difficulty of getting a representative soil sample. A more enlightened view (on pecans, but applicable to all tree and vine crops) is expressed by Darrell Sparks of Georgia:

"Regardless of the limitations of soil analysis, employment of this method is necessary if the nutrient status of the pecan grove is to be fully understood. For instance, if a soil test from a grove with zinc deficient trees indicates soil zinc is adequate, the zinc is probably unavailable to the tree and correction will most likely have to be by foliar application. If on the other hand, the test indicates zinc is low in the soil, there is a greater possibility of the deficiency being corrected by soil application. In short, while the levels of nutrients in the tree are of much greater immediate concern, soil levels can also be essential information in correcting imbalances and ascertaining possible future problems."

Sparks concludes that "in predicting the nutrient status of pecan trees, leaf analysis is superior followed by soil analysis. The nutrient status of the grove can best be predicted by using both methods in combination."

Sparks is correct. In diagnosing K deficiency, for example, the combination of leaf and soil analysis should be used. Low levels of K in leaves can be caused by stress for water - if the leaf levels of K are low and the soil levels high, water stress would be the prime suspect. Soil analysis can also help to refine interpretation of low or deficient leaf levels of other nutrients.

Summary

Soil analysis is a valuable tool if its limitations are recognized. Used in conjunction with leaf analysis it becomes a more powerful tool. Precise fertilizer recommendations cannot be made based on current soil tests but good guidelines can be established. Correlating recommendations with observed field responses for a crop will help to fine tune recommendations on a given piece of ground.

General Reference

Reisenauer, H.M. (ed.) **Soil and plant-tissue testing in California.** University of California Extension Bulletin No. 1879. April 1976.

17. PLANT ANALYSIS —
Tool Or Tribulation

Plant analysis has been so widely used and there has been so much work done on establishing "critical" levels that the value of plant analysis as a diagnostic tool has gained an exalted state that is perhaps not wholly deserved. There is no question of the value of plant analysis as a **guide** to determining nutrient deficiencies and toxicities and in planning fertilizer programs; the word "guide" should be stressed, however, as basing a recommendation for a particular nutrient solely on the plant level of that nutrient is unwise.

One of the reasons for the popularity of plant analysis is that we all like easy answers to complex questions. What could be easier than plugging an analysis figure for a particular nutrient into a chart for an instant determination of the status of that nutrient - deficient, low, sufficient, high, excessive. The wide scale publication and use of such "critical level" charts gives us confidence that this is the right approach or else why would such charts exist?

The value of plant analysis is currently coming under question in some quarters. The fertilizer industry's R.L. Luckhardt, a promoter of plant analysis through the years, has recently expressed reser-

INFLUENCE OF VARIOUS FACTORS
ON LEAF LEVELS OF SOME NUTRIENTS
- - - Nutrients Affected (leaf levels) - - -

Factor	Increases levels of:	Decreases levels of:
Moisture stress	Na, Cl, B	P, K
Irrigation	P, K	Na, Cl, B, N
Cool, cloudy weather	Nitrate	—
Heavy crop	Ca, Mg	K
Manure application	N, P, K	Zn, Mg
Potassium fertilization	K	Mg, Na
Soil salinity	Na, Cl, B	K
Soil acidity	Mn	—

vations on its value. My own field experience has taught me not to interpret plant analysis data too strictly.

A new concept called Diagnosis and Recommendation Integrated System (DRIS) is gaining acceptance in some circles. DRIS correlates a number of factors, including N/P, N/K and K/P ratios in plant tissue, to arrive at a picture of the nutrient status of a crop. The DRIS approach is believed by many to be superior to the "critical level" approach.

Some of the many factors that can influence the level of a given nutrient in a crop are given on the preceding page.

The above list is far from complete, but gives an indication of the many factors affecting the level of a particular nutrient in plant tissue. Also, to compensate for a low level of a particular nutrient, the plant can produce smaller leaves; thus the concentration in the leaf might appear normal, but the nutritional status of that element would be below optimum.

Caution should be used when interpreting leaf analysis data rather than plugging analysis results into a handy chart for an instant recommendation. If a grower intends to embark on a leaf analysis program he should know the pitfalls, or hire someome that does. Analysis data should be correlated with on-site observation of field conditions, a knowledge of crop field history and a knowledge of management practices used on the field.

Too often, shotgun analyses for all the essential elements are routinely run in the hope of finding an answer to a problem. Much of such analyses is wasteful; concentrating on the most likely possibilities is a better approach. Knowledge of molybdenum or copper levels in plants is rarely of value. On the other hand, analysis for some non-essential elements is rarely done, even though such analysis could provide useful information in some situations, e.g., aluminum and lithium analysis.

Proper plant sampling, both proper time and proper plant part, is an important part of getting reliable analysis data. Sampling guides for individual crops can be found in publications or are available from extension services. Analysis of plant tissue that has been sprayed with nutrients is of little value for those nutrients since it is difficult to remove sprayed nutrients even if leaves are washed (most labs do not do a thorough job of washing anyway); too much washing can alter results by reducing levels of soluble nutrients such as potassium.

Following is a discussion of plant analysis for three general crop groupings.

Comments on Plant Analysis for Various Crops

Row and field crops

Petiole analysis for nitrate can be very useful in adjusting N fertilizer programs on row crops. Petiole nitrate should be high early in the season then decline as the season progresses (see accompanying diagram). Sampling should be avoided during cloudy or overcast weather because of the possibility of temporary high nitrate accumulation under such conditions. Nitrate can also accumulate during very hot or very cool weather.

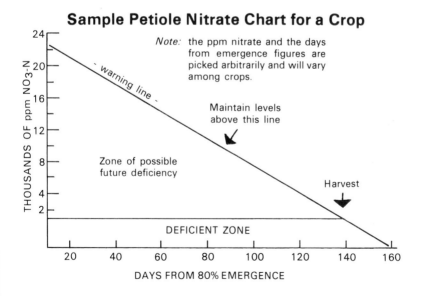

Sample Petiole Nitrate Chart for a Crop

Note: the ppm nitrate and the days from emergence figures are picked arbitrarily and will vary among crops.

Maintain levels above this line

Warning line

Zone of possible future deficiency

Harvest

DEFICIENT ZONE

THOUSANDS OF ppm NO3-N

DAYS FROM 80% EMERGENCE

At the end of the growing season it is desireable to have nitrate levels at or close to deficiency levels for earlier, more uniform harvest. The slope of the "warning line" for a crop will vary depending on soil type - the slope will be steeper for a sandy soil and gentler for a clay soil.

A petiole sampling program for nitrate must include 2 or more samples during the season - ideally, 4 or more. The nitrate levels on 2 or more samples are plotted on a graph, the points are connected, and the slope of the line is evaluated - it is the slope of the plotted line that is important and that can be used to adjust fertilizer programs during the season; the absolute value of nitrate for any individual petiole sample has much less meaning. Interpretive levels for various crops at different times during the growing season are available in publications or from extension services. Nitrate levels close to harvest can be used in deciding which of a number of fields to harvest first.

Row crop plant analysis for nutrients other than N is often not definitive because many analysis levels fall in the "gray area" between deficiency and sufficiency. Analysis for other nutrients can be useful, however, when correlated with other data (e.g., soil analysis data).

For alfalfa, P and K analysis of mid-stems can be useful. There is evidence that P/K, P/N and K/N ratios in plant tissue can be important in alfalfa.

Grapes
Growth and nutrient use by cultivated grapes are unique and totally different from any other crop. The reason for this is the severe pruning that is performed on grapes each winter. For virtually all other crops, the top:root ratio does not change significantly. For grapes, the top:root ratio is thrown way out of kilter after a heavy pruning. When starting growth in the spring, grapevines possess a relatively massive root system to supply a relatively small top. The result is vigorous growth and constantly fluctuating nutrient relationships.

Although petiole analysis for nitrate at bloomtime is often done on grapes, the results have too often been erratic and misleading. Analysis of grape **leaves** for total N, 3 or more times a year, will give a better handle on N nutrition of grapes than will petiole nitrate analysis.

Indications are that arginine analysis of canes and/or juice can give a handle on N nutrition of grapes (grapevines accumulate arginine in the fall for use the following growing season) but results have not been consistent and much more work needs to be done in this area.

Compounding the difficulty of establishing a meaningful plant analysis program for grapes is the large number of grape varieties that exist. Critical levels of nutrients are not consistent among varieties. In California, most of the experimental nutritional work has been done on Thompson Seedless; at present it is pretty much up to individual growers of other varieties to develop their own critical level data.

Orchards
Leaf analysis is widely used in orchards and has proven to be an excellent diagnostic tool. Unfortunately, however, many leaf analysis levels fall in the "gray area" between deficiency and sufficiency (this is often the case with zinc), making precise recommendations difficult. Potassium (K) levels can be very low in orchards under stress for water; one should not make a potassium fertilizer recommendation for orchards (or for any crop) based solely on leaf analysis but should get confirming data from other sources. On pecans, excessive nitrogen fertilization will induce potassium deficiency.

Caution should also be used with boron leaf levels as good response to boron has been obtained even though leaf levels of B were ample.

Leaf analysis for total calcium is virtually useless when attacking calcium stress disorders (analysis for soluble calcium may have merit in some cases); one shouldn't be lulled into thinking there are no calcium problems just because leaf Ca is high.

Like grapes, much of the spring growth of trees comes from nitrogen that has been stored during the fall and winter. Bark and wood tests during the winter for arginine are being tried for orchards and may prove to be of benefit in the future.

Detecting and Monitoring Toxicities

An excellent use of leaf analysis is in the detection and monitoring of toxicities. Leaf analysis is probably more useful in diagnosing toxicities than it is in diagnosing deficiencies.

Monitoring sodium, chloride and boron in leaves can be helpful in detecting soil salinity build-up. This can be especially valuable in drip irrigation where salts build up on the periphery of the wetted zone. Currently, it is difficult to measure the soil salinity build-up from drip irrigation so that a flushing irrigation can be applied, if needed. Using leaf analysis, the drip irrigated plant can be looked at as a biological salinity sensor giving readouts on soil salinity in terms of sodium, chloride and boron levels in the leaves.

Summary

Plant analysis can be a useful tool, an extremely useful tool, in diagnosing deficiencies or toxicities and in planning fertilizer programs. It is not, however, a definitive tool. Plant analysis data should be correlated with soil analysis data, water analysis data, soil type and irrigation data, crop and field history data, **and field** observations. Only by doing so will the maximum benefits be obtained from plant analysis.

18. FOLIAR FERTILIZATION

Foliar nutrient sprays have been used in agriculture for a long time. If they did not have merit, they would not have lasted this long.

Foliar sprays are used for 3 main purposes:

1. **To correct a deficiency**
   When a deficiency of a particular nutrient is spotted in the field, waiting for a soil application to take effect can mean crop dollars lost. Foliar sprays can give fast correction of nutrient deficiencies; zinc sprays have been widely used in this regard.

2. **To maintain optimum nutrition of a particular nutrient**
   Vigorously growing plants can grow faster than the ability of their roots to supply the tops and fruit with a needed nutrient. Zinc deficiency on vigorously growing young almond trees is an example. One or more zinc sprays during the growing season can help to maintain optimum zinc nutrition of crops.

   Calcium sprays are used to prevent disorders on crops that do not have the ability to supply calcium to fruiting parts that have a need for calcium.

   When soil temperatures are unseasonably cool, as can occur during some springs, the adsorbtive efficiency of a plant's root system is impaired. Foliar nutrient sprays can help plants through such periods.

3. **To give a crop a nutritional boost at a critical juncture in it's life**
   On tree fruit crops, a critical growth stage occurs immediately after pollination when a large number of small fruits (or nuts) compete for a limited amount of nutrients; those that don't survive the competition drop. On other crops, the period of seed set and pod filling is a critical one. For cotton, the period of heavy boll set is critical. Providing

a nutritional boost with a foliar spray can help crops make it through these critical periods with minimal crop loss.

On citrus in California, a urea-zinc spray near the time of fruit set has become an accepted and University recommended practice.

Fall-applied boron sprays on prunes can strengthen next years developing flowers so that they set much better the following spring; this effect occurs even though boron levels in soil and leaves are in an optimum range.

Recent work in soybeans in Iowa has shown increased seed set from foliar nutrient sprays.

Mechanism of Foliar Absorbtion

There are 2 main paths of entry for foliar nutrients: the cuticle and the stomata. These are discussed separately:

Cuticle entry

The cuticle is a waxy layer covering leaves, flower petals and fruits. It's primary function is to prevent moisture loss from plants and to protect the plant from injury. It is typically thicker on upper leaf surfaces than on lower leaf surfaces and is impermeable to aqueous (water) solutions; oily solutions will penetrate the cuticle more readily. Cracks can occur in the cuticle and can serve as a port of entry for foliar sprays. On young, immature leaves the cuticular layer is not as well developed as an older leaves.

Stomatal Entry

The stomata of plants are located almost exclusively on the leaves. These pores are only a few millionths of an inch in diameter and their primary function is to breathe for the plant. They allow carbon dioxide, the building block of plants, to pass into the leaf and here the chlorophyll molecules, using sunlight for energy, convert the carbon dioxide molecules into simple sugars. The stomata also allows the by-product of this reaction, oxygen, to escape from the plant. A third function of the stomata is to allow the escape of water vapor that evaporates from the inside of the leaves to cool the plant.

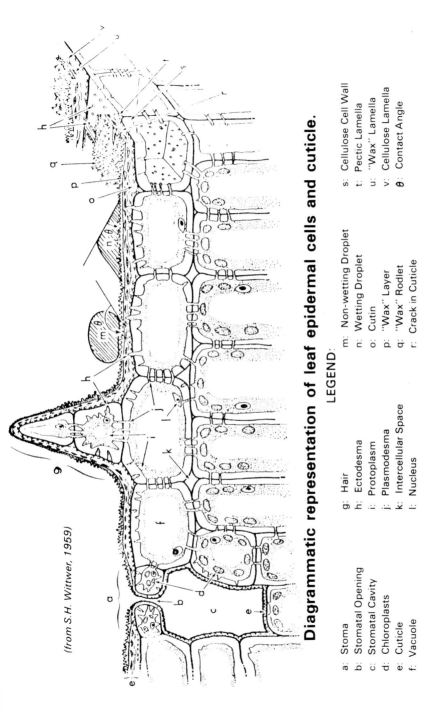

(from S.H. Wittwer, 1959)

Diagrammatic representation of leaf epidermal cells and cuticle.

LEGEND:

a: Stoma
b: Stomatal Opening
c: Stomatal Cavity
d: Chloroplasts
e: Cuticle
f: Vacuole

g: Hair
h: Ectodesma
i: Protoplasm
j: Plasmodesma
k: Intercellular Space
l: Nucleus

m: Non-wetting Droplet
n: Wetting Droplet
o: Cutin
p: "Wax" Layer
q: "Wax" Rodlet
r: Crack in Cuticle

s: Cellulose Cell Wall
t: Pectic Lamella
u: "Wax" Lamella
v: Cellulose Lamella
θ: Contact Angle

Stomata are usually closed at night, and can close during the hottest part of the day.

The distribution and occurrence of stomata, as well as their size and shape, varies widely from one plant species to another. Broadleaf crops and tree crops commonly have most of their stomata on the lower leaf surface while grass species may have about the same number on both surfaces. As examples, small white beans commonly have 40,000 stomata per square centimeter on their lower surface and only 3,000 on their upper surface, while sorghum has 16,000 on the lower surface and 11,000 on the upper surface. In addition, the sorghum stomata are four times as large as the bean stomata.

The accompanying diagram depicts the cuticle and a stoma of a leaf.

Timing of Foliar Sprays

Timing of foliar sprays can be critical, both in regard to time of season and time of day:

Optimum time of season

As indicated previously, foliar sprays may be effective only during "critical stages" of a plant's growth cycle and must be applied during or shortly before the critical period to be effective. For example, zinc deficiency on grapes causes shattering (berry drop) of clusters; pre-bloom foliar zinc sprays can prevent shattering, post bloom sprays will not.

Since immature foliage does not have a well developed cuticular layer, application of nutrient sprays when there is a significant amount of young foliage present will enhance cuticular entry.

Optimum time of day

For maximum stomatal entry, nutrient sprays must be applied when the stomata are open. Since stomata can close at night and during mid-day, early morning applications are the best. Also, there is less evaporation during the early morning thus giving a better chance for maximum uptake by leaves. High relative humidity during the time of application will also enhance uptake by minimizing evaporation.

Methods of Application

For crops having most of their stomata on the underside of leaves (most broadleaf and tree crops) ground rig application is preferred. For crops having significant stomata on the upper leaf surface (most grasses) either ground or aerial application is satisfactory.

Surfactants

Use of surfactants or wetting agents is important to get maximum benefit from foliar applied nutrients. Surfactants lower the contact angle of spray droplets on the leaves (see diagram) thus enhancing absorbtion.

New chemicals are being tested that have the ability of penetrating or breaking down the waxy cuticular layer of leaves. Advances here would greatly widen the scope of foliar fertilization.

Individual Nutrient Sprays

Individual nutrient sprays are discussed in four categories following:

Nitrogen sprays

When foliar nitrogen is needed, urea is the material usually selected. Other material can be used, but have an increased phytotoxicity hazard except at very low rates. Tolerance of various plants species to foliar applied urea is given in the following table:

Tolerances of plant foliage to repeated applications of urea sprays [†]	
CROP	Tolerances—no leaf scorch* (Lbs/100 gal. water)
VEGETABLE	
Cucumber, bean, tomato, sweet corn	4-6
Carrots, celery, onions, potatoes	15-20
FRUIT	
Apple, grape, brambles, strawberry	4-6
Cherry, peach, plum, currants	10-20**
FIELD CROPS	
Sugar beets, alfalfa, corn, wheat	5-10
Oats, barley, bromegrass, ryegrass	20-50
PLANTATION AND TROPICAL	
Tobacco, citrus, cacao	5-10
Sugarcane, banana, mango, tea, coffee	10-25
Cotton, hops, pineapple	20-50

* Higher concentrations may be used with lower volume spraying.
** Leaves are tolerant at this level as measured by scorch, but severe chlorosis may occur at these and even lower concentrations.

[†] from **Foliar Application of Nutrients** by S.H. Wittwer. Plant Food Review, No. 2, 1967.

Biuret is an impurity in many urea formulations and is toxic to many plant species at high concentrations. Urea with a low biuret content is available (Lo-Bi urea) and should be used for foliar sprays.

When applied in conjunction with zinc, urea can enhance the uptake of zinc by leaves. Zinc-urea combinations are common for foliar application.

Application of urea will often "green up" a plant (as can some other nutrients). This greening does not always mean that an economic yield increase has (or will) occurred. Crop yield measurements are the best method of evaluating the efficacy of a foliar spray.

Urea sprays have not been effective on stone fruits.

Other major nutrient (or macronutrient) sprays
Phosphorus (P) and potassium (K) are not commonly applied as foliar sprays but have been used successfully on some crops. Heavy cropping of prunes creates a temporary potassium deficiency that is alleviated by potassium nitrate sprays at bi-weekly intervals.

Calcium sprays are often the only method of correcting calcium related disorders. Both calcium nitrate and calcium chloride are used to correct bitter pit in apple.

Magnesium sprays, usually magnesium sulfate, are not common but have been successfully applied. Sulfur sprays are rare.

Micronutrient sprays
One of the first uses of foliar sprays was in the correction of zinc deficiency. Foliar zinc sprays are widely used today and will be in the future. Deficiencies of the other metallic micronutrients (copper, iron, manganese) can also be alleviated by foliar sprays. Both inorganic and chelated materials have been used for correction. Chelates, though more costly, have outperformed the inorganic materials is some tests.

Boron sprays are common on orchard crops in the pacific northwest and are also used on crops in other areas. Molybdnum sprays have been used to correct Mo deficiency.

Carbohydrate sprays

Although not always considered as such, carbon, hydrogen and oxygen are major, essential plant nutrients. Fruit drop at the critical fruit setting stage on some crops is felt by some to be related to carbohydrate shortage (the definition of carbohydrate is a material that contains carbon, hydrogen and oxygen). Some cotton growers feel that sugar sprays during the bloom period increase boll set, but there is no hard data on this. Some greenhouse growers have shown definite benefits from CO_2 enrichment of greenhouses.

Interestingly, urea is a carbohydrate-formula $CO(NH_2)_2$. Probably most of the carbon in urea is converted to CO_2 shortly after being sprayed but detailed studies on the fate of the carbon in urea would be of interest.

Summary

Foliar nutrient sprays have a definite place in agriculture and will see more use in the future. Fieldmen should be alert to the possibility of mixing foliar nutrients with insecticides and thus giving them a free ride. Where fruit or crop spotting is a concern, obtain experience with a particular nutrient formulation before using it on a large scale.

Shot gun mixes of nutrient sprays should be avoided; the rifle approach is better. Discover what nutrient is the most limiting and concentrate on it. Beware of the many foliar nutrient formulations on the market. Many of them carry a fancy price tag but are little, if any, better than the cheaper inorganic, non-proprietary materials.

General Reference

Wittwer, S.H. and F.G. Teubner. **Foliar absorption of mineral nitrients.** Annual Rev. of Plant Physiol. 10:13-32. 1959.

19. ORGANIC AND BIOLOGICAL FERTILIZERS AND AMMENDMENTS

The word "organic" has a mystical quality about it - many associate the word with purity and plain old fashioned goodness. In contrast, "inorganic" can seem a cold, sterile word and "chemical" far worse.

Plant scientists in the nineteen thirties taught us that inorganic chemicals could provide complete nutrition of crops. Today, commercial hydroponic operations (where all nutrients are supplied in an inorganic form to a water solution) produce a variety of high quality vegetables (ironically, some of this hydroponic produce is sold as "organic" produce). It has been only relatively recently that agriculture has used inorganic fertilizers in quantity. Today, inorganic fertilizers are used, usually exclusively, to supply the nutrient needs of crops on virtually every agricultural operation in the U.S.

Organic fertilization in this country goes back to when the Indians taught the pilgrims that planting fish with corn was beneficial to the corn. Later, manure was widely used as a nutrient source. The popularity and widespread use of inorganic fertilizers today is due to their high analysis of essential nutrients. Crops could still be successfully grown today solely with manure and other organic materials but the relatively low nutrient analysis of such materials means that it's just not economical to transport them around the country. (Manure is still an economical and valuable fertilizer material when hauling distances are short and economical. Use of sewage and industrial wastes in agriculture is increasing and is as much a disposal problem as an agricultural use consideration; the benefits of disposal tilt the normal cost:benefit calculations toward the use of such materials).

Part of the mystique of organic fertilizers lies in their appeal to the sense of longing in many of us to return to what is perceived as simpler and easier times. For the unknowledgeable, the idea that inorganic fertilizers are sterilizing the soil is a tempting concept.

Every year, salesmen for some "new" organic or biological material make their appeal to these feelings. The above is not to say that there is no benefit from applying organic materials - it's just that in most cases the costs far exceed the benefits.

A major benefit or organic additions to soils is on the physical condition of the soil (rather than a nutritional benefit). Organic matter addition can result in improved tilth and permeability of many soils. Unfortunately the effect is only temporary and again, the costs may exceed the benefits.

Soils that would benefit most from organic ammendments are, logically, those that are very low in organic matter to start with. A main reason that soils are low in organic matter it that they occur in hot, arid areas. Heat and moisture greatly accelerate the decomposition of organic matter and as a result it is difficult to maintain organic matter levels on irrigated lands in hot, arid areas. Any organic material added will be effective only a short period of time during a hot growing season (effects will be longer during the cool part of the year, but the effects are usually desired during the warm part of the year). Like a newly hatched bird, low organic matter soils in hot, irrigated areas have an insatiable appetite - an appetite for organic matter. An exception to the preceding is some orchards, where shading of the ground by trees reduces soil temperature to the point where organic matter can build up. Judicious use of mulches and organic materials in such cases may be helpful in improving soil tilth and permeability.

When considering a proposed organic ammendment, the farmer (or fieldman) should test it on a small plot before any large scale application is considered. Better yet, wait for a thorough, unbiased test by University personnel - positive results from such tests are extremely rare; when they do occur the economics are usually prohibitive. **Don't** be influenced by testimonials.

Beware of biological ammendments that "bring a soil to life". They might work in a test tube, but in the relatively hostile environment of agricultural soil they are unlikely to maintain sufficient populations to be of any benefit.

Current University work on biological additives, including bacteria and algae, has shown promise in pot tests but results have not yet been translated to field tests. Mycorrhizae, a common root fungus, has shown significant benefits, including nutritional benefits, in pot and small plot tests; again, field scale results have not been obtained.

There may well be a day when some of these materials will be successful in the field. In the meantime, the farmer-fieldman should look to Universities rather than salesmen for solid information on organic or biological materials.

General References

Allison, F.E. **Soil organic matter and its role in crop production.** (Developments in Soil Science, Vol. 3), Elsevier Publishing Co., 52 Vanderbilt Ave., N.Y., N.Y. 10017. 1973. (A 637 page book).

Dunigan, E.P. And J.F. Parr. **Soil activators and conditioners, an unlikely panacea.** Crops & Soils, April-May 1975. p. 12-15.

Stevenson, F.J. **Humates - facts and fantasies on their value as commercial soil ammendments.** Crops & Soils, April-May 1979. p. 14-16.

20. THE FUTURE OF SOIL AND PLANT NUTRITION

The future potential benefits from improved soil and plant nutrition are virtually limitless. This is in contrast to the benefits to be obtained by improving other aspects of crop production. For example, there is a finite limit to obtaining improved yields through improved plant spacing; improved yields through pest and disease control are limited by the fact that you can only kill a particular pest so dead.

New higher yielding varieties and improved farm management will continually exert pressure on improvement of plant nutrition. Higher per acre yields from improved varieties and improved farm management are a given. Higher yields, by definition, mean higher nutrient use and demand by crops. Couple this with the wider use of purer, "nutrient free" water for irrigation and with longer and longer individual field histories of nutrient removal and it becomes clear that soil and plant nutrition will be an exciting and active field of endeavor for many years to come.

Two major areas for improvement in soil and plant nutrition are old ones:

1. **Improved soil and plant analysis techniques** - faster and more accurate methods.

2. **Improved interpretation of soil and plant analysis data** - e.g., the DRIS method of interpreting plant analysis data.

These two areas always have been and always will be areas for improvement.

Many problems in plant nutrition lay in between the disciplines of plant science (or horticulture) and soil science - the calcium stress problem is an excellent example. A cooperative, interdisciplinary approach to solving such problems is the best approach.

The fieldman should never accept current concepts in soil and plant nutrition without some reservations. All concepts are subject to change and improvement. Two tenets that were among those most drilled into me during my college training were:

1. Soil tests for nitrogen, including nitrate, are a poor way to evaluate the nitrogen needs of row crops because of the transient nature of soil N.

2. Deciduous orchards (and vineyards) should be fertilized in the winter so that nutrients are in place when the tree makes it's growth surge in the spring.

The reasoning behind these tenets was logical and well presented, however both tenets have fallen. Soil nitrate tests can provide good information for row crops. It has been found that a deciduous tree's growth surge in the spring comes mainly from nitrogen materials stored during the late summer and fall and that late summer or fall fertilization is superior to winter fertilization.

We have a tendency to accept things as being true if they are printed in black and white, more so, if the printing and paper quality are good, more so yet if they are found in an authoritative publication, even more so if they come from an authoritative source. **Nothing** should be accepted without question.

It is inevitable that some of the presently accepted tenets in soil and plant nutrition will fall by the wayside in the future to be replaced by sounder tenets which in turn will be subject to refinement and replacement. This process is a never ending one and should not be resisted - it is called progress.

By definition, the resistance to a new idea can impede progress. There is undue pressure, especially in the U.S. on being correct or on not being wrong - people are too often judged by their mistakes rather than by their accomplishments. The result is the inhibition of new, potentially worthwile ideas. This is especially true in academic agriculture where the penalties for being wrong are excessive. It is a virtual impossibility to be correct 100% of the time - being correct 60% of the time on 1000 efforts can be superior to being correct 100% of the time on 1 effort, (if one is thoroughly conscious of the fact that no idea is infallible); productivity is more important than batting average.

Often the discussion of an incorrect idea can be more meaningful and can bring us a bit closer to the truth than the enlightenment engendered by a correct idea. Albert Einstein was well aware of this. Einstein was always interested in honing, discarding, reworking and changing his theories, forever searching for a more refined, a purer truth. He was more interested in getting at the truth than in whether his own particular ideas were correct. He was willing, even anxious, in inviting criticism of his ideas and had no qualms about abandoning an idea if it's lack of merit could be demonstrated; he was, in fact, delighted when someone could point out a mistake or improve on one of his ideas. Einstein was, and is, revered by those in his fraternity as much for this latter quality as for his scientific accomplishments. The combination of both is a powerful one - a case where the whole is greater than the sum of the parts.

Too often we are concerned solely with being right. The merits of positive, meaningful, correct results cannot be denied, but with the pressure for results, a part of the imagination and capability of man is sacrificed. Academic (and some other) institutions can discourage rather than encourage original thinking. In a thoughtful piece, one agricultural scientist offered these views:

"It is a foolish young scientist who does not soon learn that the Court of the Inquisition still sits in judgement on the unorthodox. In 1616, it forced Galileo to submit to the orthodoxy of the theologians. Today's equivalent comes from a web of bureaucratic rules and policies. All are well intentioned, but their cumulative effect is to exert a powerful pressure towards orthodoxy.....Our bright young people need to be allowed a little more freedom and a chance to prove themselves original thinkers without the system whipping them into traditional, sometimes unimaginative channels."

The future of soil and plant nutrition depends a great deal on the quality of people that enter into the field. In the U.S., there is currently more than the usual emphasis on making as much money as possible in life. As a result, some of the better young minds in the country are attracted to fields where they feel the money is. For some such minds, this might mean a career as a lawyer rather than a career in agriculture. A recent rebuttal to this approach makes a good point:

> "The real trouble with the American legal system is the amount of society's resources it diverts from truly creative and productive activity. We see this on the level of the individual lawsuit, and we see it in whole industries paralyzed by overly complex and often illogical restraints. It is hardest to see, but most damaging, on the level of the individual lawyer. As Carter pointed out, the U.S. has three times as many lawyers per capita as England and 21 times as many as Japan. Year by year our system of social rewards entices many of our brightest and most energetic young adults into this essentially fruitless activity as their life's work. How many scientific breakthroughs and great novels and management innovations have we lost by this absurd arrangement?"

The preceding is overstated as there are many in the legal profession engaged in useful, productive work; it does, however, make a point.

There are a significant number of excellent, brilliant minds in the field of soil and plant nutrition and in the field of agriculture as well, men and women that would have been highly successful and far better off financially in other lines of work. They are aware however, of the rewards to be found in useful work. To attract the best minds into soil and plant nutrition, the appeal must be made to the value of doing useful work rather than the monetary rewards.

For those that might doubt the meaningfulness, or, to use a word currently in favor, relevancy, of plant nutrition or agriculture as a career, consider these words from a leading agriculturist prior to the 1977 meeting of the American Society of Agronomy:

"Probably, no meeting in 1977 of politicians, bureaucrats, social reformers, urban renewers, modern-day Jacobins, or anarchists will cause as much change in the economic and social structure of the country as the ASA meeting of crop and soil scientists."

Overstated? Perhaps; then again, maybe not.

General Reference
Vallery-Radot, Rene. **Life of Pasteur.** Garden City Publishing Co., Inc., Garden City, N.Y. (N.D., circa 1923).

— REFERENCE NOTES —
CHAPTER 1
Note: Omit tables and headings when counting lines

page 5, bottom

Related to this observation is the salt tolerance of experimental grain varieties being irrigated by sea water in tests near Bodega Bay, Ca. Some varieties show excellent salt tolerance at the on-site location, which has a cool, foggy, climate. The salt tolerance of these varieties will probably not be as good in a hot climate.

page 6, line 6

Bradford, Gordon. **Lithium in California's water resources,** Calif. Agric. May 1963, p 6.

page 6, lines 7,8

See General Reference No. 1 at end of chapter for tables showing guideline levels for various trace elements. These guidelines will be firmed up with time.

page 7, SAR

See Gen. Ref. No. 5 at end of chapter.

page 7, RSC

See Gen. Ref. No. 2, Eaton's paper on carbonates is an intriguing blend of scientific and literary writing that could qualify as a work of art in the field of literature as well as science - quite different from the "just the facts ma'am" papers found in today's scientific journals.

page 8, lines 5-15

Gen. Ref. No. 3

page 8, adj SAR

Rhoades, J.D. **Quality of water for irrigation.** Soil Sci. 113:277-284. 1972.

adj. SAR may undergo further adjustment as time goes by. One proposed adjustment was given in 1979, (Oster, J.D. and F.W. Schroer, **Infiltration as influenced by irrigation water quality.** Soil Sci. Soc. of Amer. 43:444-447. 1979).

page 9, 10, pHc

Tables for calculating pHc were taken from Gen. Ref. No. 4. The other lime depositing index mentioned (the Langelier Index) goes back to 1936 (Langelier, W.F.. **The analytical control of anti-corrosion water treatment.** J. Amer. Water Works Assoc. 28:1500-1521. 1936).

page 12, bottom

Pratt, P.F., R.L. Branson and H.D. Chapman. **Effect of crop, fertilizer, and leaching on carbonate precipitation and sodium accumulation in soil irrigated with water containing bicarbonate.** 7th Intern. Cong. of Soil Sci., Madison, Wisc., U.S.A. 1960.

page 13, lines 12-15

See Gen. Ref. 3

page 13, lines 23-25

Another report of lime accumulation on plant roots comes from Wyoming. On a crop of sainfoin, 57% of dry weight of roots in one test was calcium carbonate. (Ross, W.D. and R.H. Delaney. **Massive accumulation of calcium carbonate and its relation to nitrogen fixation of sainfoin.** Agron. J. 669:242-246. 1977). Because we normally only observe the above ground portion of plants we tend to forget about what might be happening below the ground. It would be nice if we could see the root system each time we examine a field.

page 14, lines 1-14

See Gen. Ref. 1

page 17, lines 26-29

Dow, A.I. **S fertilization of irrigated soils in Washington State.** Sulphur Institute Journal, Spring 1976, p 13-15.

CHAPTER 2

page 21, line 1

Personal communication

page 21, lines 4,5

Olsen, S.R. and R.S. Watanabe. **Solubility of calcium carbonate in calcareous soils.** Soil Sci. 88:123-129. 1959. This paper shows how the bicarbonate concentration of the soil solution can vary among soils.

page 21, table

Hassen, M.N. and R. Overstreet. **Elongation of seedlings as a biological test of alkali soils. I. Effects of ions on elongation.** Soil Sci., 73:315-326. 1952.

CHAPTER 3

page 23, lines 4,5

Most of these salinity scientists headquartered at the USDA Salinity Lab in Riverside, Ca. This lab was established in 1937 and is the source of most of the past and present information on saline-alkali soils.

page 23, bottom

Gen. Ref. 1, pp 18, 30

page 24, lines 4,5

Gen. Ref. 1, p 30

page 24, lines 13-16

Bower, C.A., personal communication, 1969; see also, Method 3b, p 88, in Gen. Ref. 1

page 24, line 21

Considerable thought should be given to step 1 (grading the land). Salts and boron usually increase with soil depth and on some soil series, significant cuts will expose soil that is very poor from both a physical and chemical standpoint and that would take many years to reclaim; sprinkler or drip irrigation (with minimal grading) would be superior to surface irrigation in such cases. Where practical, level-basin irrigation should be considered; see **Level basin irrigation: a method for conserving water and labor.** USDA Farmers' Bulletin No. 2261, April 1979.

page 26, lines 3,4

Gen. Ref. 2 and also found in other literature. 1 acre foot of water will remove significantly less than 80% of the salts from a foot of soil if subsoil restrictions occur; the 80% figure assumes the soil has good draingage.

CHAPTER 4

page 29, lines 13-15

Overstreet, Roy, J.C. Martin and H.M. King. **Gypsum, sulfur and sulfuric acid for reclaiming an alkali soil of the Fresno series,** Hilgardia. 21 (5):113-127. Nov. 1951.

In the above study, sulfuric acid was definitely superior to both gypsum and sulfur (on an equivalent basis) under rigid experimental test conditions. Some growers have a preference for sulfuric acid over other ammendments because they have observed its superiority first hand.

San Joaquin Valley row crop farmers successfully apply sulfuric acid to the surface of the soil and incorporate it only a few inches; followed by sprinkler irrigation, this practice can provide a good environment for the sensitive seed germination and seedling growth stages on some soils. With any ammendment, the entire root zone does not have to be reclaimed as plants can make good growth when only a portion of the root zone is reclaimed; see Lunin, J. and M. Gallatin, **Zonal salinization of the root system in relation to plant growth.** Soil Sci. Soc. of Amer. Proc. 29:608-615. 1965. This concept correlates with the effectiveness of drip irrigation on highly saline soils and with the ability of boron sensitive tree crops to do well on soil that has very high boron levels, but only in part of the root zone.

page 30, lines 19-22

Although a good ammendment, ammonium polysulfide (Nitro-Sul) is highly corrosive and therefore not widely used. On the other hand, ammonium thiosulfate (Thio-Sul) is relatively non corrosive and is gaining wide acceptance; with an N content of 10 to 12% it is a combination ammendment-fertilizer.

page 30, lines 25-27

When using a combination of lime on the soil + acidifying ammendments in the water, it should be remembered that volatilization losses of nitrogen (to NH_3) are much greater when there is lime in the topsoil. Incorporation and/or immediate watering in of nitrogen fertilizers that contain urea or the ammonium form of nitrogen are much more important when there is lime in the topsoil.

page 30, lines 31,32

Miyamoto, S., J. Ryan and J.L. Stroehlein, **Sulfuric acid for the treatment of ammoniated irrigation water: I. Reducing ammonia volatilization.** Soil Sci. Soc. Amer. Proc. 39:544-548. 1975.

page 31, lines 3-5

The largest manufacturer and distributor of SO_2 generators is the D & J Harmon Co., Inc., 3737 Gilmore Ave., Bakersfield, CA 93308.

page 31, bottom
Hudson, Bill. **Yakima Valley Fruit Facts.** The Goodfruit Grower, March 15, 1974, p 12. Hudson gives an excellent rundown on the history and uses of lime sulfur in orchards in the pacific northwest.

CHAPTER 5

page 35, bottom
The inclusion of sodium nitrate as a possible nitrogen fertilizer may offend some, but this material can have a place in certain situations. When a quick N boost is needed on row crops, sodium nitrate (or nitrate of soda) can be very effective, and not harmful for many soils. A 1978 article praises, probably over-praises, the benefits of sodium nitrate on cotton (**A good cotton fertilizer.** American Cotton Grower, April 1978, p 38, 47.) See also the end of Chapter 1 for a discussion of sodium nitrate.

page 36, lines 5-7
The conversion or hydrolysis of urea to ammonium is usually referred to as a rapid process, however the term "rapid" is imprecise. Dr. P.G. Moe has caculated that the "conversion which normally takes place in about one week in the spring when the soil temperature is about 70^0 can require about one month when the soil temperature is down around the freezing point, or, conversely, be accomplished in only one day in the summer when soil temperatures might be around 100^0F." (paper presented at Farm Science Days, Purdue, Univ., Jan. 20, 1964; quoted in Collier Co. Urea Manual for Fieldmen by R.L. Luckhardt., 1977).

page 36, table
Aldrich, Tom. **Nitrogen, when and how to use it.** Diamond/Sunsweet News, April 1979, p 13.

The volatilization table was taken from the above reference. A more detailed discussion of N volatilization losses is given in: Sharratt, W.J., **Nitrogen Efficiency, the far west.** Solutions, Sept.-Oct. 1975. p 24, 26, 28, 30.

Much of the basic research on which the above reports are based was done by Dr. L.B. Fenn of Texas A & M.

page 37, first table
Acidifying properties of N fertilizers were taken from

Western Fertilizer Handbook. (see General References); the acid-lime calculations are open to question since different workers have come up with different answers; see Tisdale and Nelson (Gen. Ref.) for a different computation of acidifying potential.

page 37, second table
The sulfuric acid-soil pH table is rough. Both this chart and the preceding fertilizer-acidity chart should not be interpreted precisely. They are rough approximations only; they can, however, give a good indication when pH problems might arise on a particular piece of ground that is under a certain fertilizer regime. Supplemental pH testing of soil will give a more accurate picture.

page 38, line 24
Sparks, Darrell. **Nitrogen scorch and the pecan.** Pecan South, Sept. 1976. 3:500-501, 1976.

page 39, nitrite
Paul, J.L. and E. Polle. **Nitrite accumulation related to lettuce growth in a slightly alkaline soil.** Soil Sci. 100:292-297. 1965.

CHAPTER 6

page 42, lines 12-14
See first Gen. Ref. at end of chapter

page 43, lines 19-21
Clement, Lawrence. **Sulphur increases availability of phosphorus in calcareous soils.** Sulphur in Agriculture (a publication of The Sulphur Institute, 1725 K St. NW, Washington, D.C. 20006) Vol. 2, p 9-12. 1978.

John Ryan and J.L. Stroehlien. **Use of sulfuric acid on phosphorus deficient Arizona soils.** Progressive Agriculture in Arizona (U. of Ariz.) 25(6):11-13. Nov.-Dec. 1973.

John Ryan and J.L. Stroehlein. **Sulfuric acid treatment of calcareous soils: Effects on phosphorus solubility, inorganic phosphorus forms, and plant growth.** Soil Sci. Soc. Amer. J. 43:731-735. 1979.

CHAPTER 7

page 45, lines 12,13
Nelson, W.L. **Potassium: one key to high quality, high yield crops.** Solutions, Nov.-Dec. 1975, p 64, 66, 68, 70, 71, 72. (Iowa, N. Dakota reference).

page 45, lines 13,14
Skogley, E. **Potassium soil test recommendations not always accurate.** Crops & Soils. Oct. 1976, p 15-16. (Montana reference).

page 45, lines 14-18
Studies on the potassium supplying power of soils. The Deciduous Fruit Grower, March 1980, p 78.

page 45, lines 18-20
Brown, A.L., J. Quick and G.J. DeBoer. **Diagnosing potassium deficiency by soil analysis.** Calif. Agric. June 1973, p 13-14.

page 45, lines 26-32
Wilcox, G.E. and R. Coffman. **Simplifying plant evaluation of K status.** Better Crops with Plant Food, Spring 1972. p. 9, 30.

page 45, lines 33,34
McLellan, G.W. **Potassium's spectacular economic benefits.** Agrichemical West, Oct. 1968. p 6, 8, 10, 12, 14, 16.

page 46, lines 17-19
Carlson, R.M. et al. **Displacement of fertilizer potassium in soil columns with gypsum.** J. Amer. Soc. Hort. Sci. 99:221-222. 1974. K-Ca interactions have been reported (one can depress the plant uptake of the other); the displacement benefits of Ca could therefore be offset in some cases.

page 46, lines 21-23
Uriu, K., et al. **Potassium fertilization of prune trees under drip irrigation.** J. Amer. Soc. Hort. Sci. 105:508-510. 1980. Significant downward and lateral movement also occurred when potassium was applied directly through the drippers. The calcium content of the irrigation water was not mentioned in this study; it should have been.

106

page 47, lines 13,14
Stebbins, R.L. **Goodfruit growing in Oregon.** The Goodfruit Grower, Aug. 15, 1973, p. 6.

Studies on the potassium supplying power of soils. The Deciduous Fruit Grower, March 1980, p 78.

CHAPTER 8

page 48, line 8
Martin, W.E. **How to predict fertilizer needs of alfalfa.** Univ. of Calif. Ag Extension mimio, Dept. of Soil Science. (also reprinted in Kern County (Ca.) Ag Extension newsletter of March 17, 1972). 1972.

page 48, lines 8-10
Olson, R.A., (ed.). **Fertilizer Technology and Use** (2nd ed.) p 337. 1971.

Page 48, lines 15-16
Chapman, Homer (ed.) **Diagnostic criteria for plants and soils.** p 448. 1966.

page 48, lines 16-18
Christensen, N.W. and P.O. Kresge. **Sulphur response in Montana.** Sulphur in Agriculture (published by The Sulphur Institute). 3:2-3. 1979.

page 49, lines 18,19
Brezonik, P.L. et al. **Acid precipitation and sulfate deposition in Florida.** Science 208:1027-1029, May 1980.

The above reference describes monitoring of S in rainfall at 24 sites in Florida. An estimated 7 lbs of S/acre per year was added by rainfall in some areas. The deposition of S, if any, in fog would be more difficult to measure.

page 49, lines 24-27
Reisenauer, H.M. **Sulfur needs of western states.** Agrichemical West. October 1961.

page 49, lines 27-30
Dow, A.I. **S fertilization of irrigated soils in Washington state.** Sulphur Institute Journal. Spring 1976. p 13-15.

It could be argued that a sulfur content of 5 lbs per acre foot does not meet the sulfur needs of a crop such as alfalfa. This is true, and a sulfur response is certainly possible with a water source containing 5 ppm S or more. The farmer, however, that would put all the nutrients on a crop in the proportion and amounts they were used by that crop would soon go broke.

page 49, bottom
Meyer, R.D. and D. Marcum. **Alfalfa response to rate and source of sulfur.** Soil and Water (publication of Univ. of Calif. Dept. of Soil and Water Science), #43, Winter-Spring 1979-1980. This is an excellent report on sulfur deficiency of irrigated alfalfa but the S content of the irrigation water is not even mentioned.

CHAPTER 9

page 51, lines 2-4
Not as big a statement as might appear at first glance. Already farms in many areas use more calcium than any other nutrient, mainly in the form of lime or gypsum. A well with a moderate calcium content can add several hundred pounds of calcium per acre annually.

page 51, lines 6-8
See Gen. Ref. 1 (at end of chapter)

page 51, lines 11-12
See Gen. Ref. 2

page 51, lines 28, 29
Epstein, Emanuel. **Mineral Nutrition of Plants: Principles and Perspectives.** pp 306-308. John Wiley & Sons, New York. 1971.

page 51, lines 29,30
Ibid.

page 51, line 30, page 52, line 1
The ameliorating effect of calcium on boron toxicity is covered in the following 2 references: Cooper, W.C., et al. **Response of grapefruit on two rootstocks to calcium additions to high-sodium, boron-contaminated, and saline irrigation water.** Soil Sci. 86:180-189. 1958. and Reeve, E. and J.W. Shive **Potassium-boron and calcium-boron relationships in plant nutrition.** Soil Sci. 57:1-14. 1944.

page 52, lines 1-5
Epstein, Emanuel (see above)

page 53, lines 1-3
Gerard, C.J. and Hipp, B.W. **Blossom end rot of "Chico", and "Chico Grande" Tomatoes.** Proc. Amer. Soc. Hort. Sci. 93:521-531. 1968.

page 53, lines 6-8
Epstein, (see above)

page 53, lines 8-10
Raabe, R.D. and J. Vlamis. **Rooting failure of chrysanthemum cuttings resulting from excess sodium or potassium.** Phytopathology 56:713-717. 1966.

page 53, lines 10-12
Bower, C.A. and L.M. Turk. **Calcium and magnesium deficiencies in alkali soils.** J. Amer. Soc. Agron. 38:723-727. 1946.

page 53, lines 12-14
Geraldson, C.M. **Evaluation of the nutrient intensity and balance system of soil testing.** Proc. Soil Crop Sci. Soc. Florida 27:59-67. 1967 and Geraldson, C.M. **Intensity and balance concept as an approach to optimal production.** p. 353-364. **In** R.M. Samish (ed.) Recent Advances in Plant Nutrition. Gordon and Breach Publishers, N.Y. 1971.

page 53, lines 26,27
Bower, C.A. and L.M. Turk (see above)

page 54, lines 1,2
Spurr A.R. **Anatomical aspects of blossom-end rot in the tomato with special reference to calcium nutrition.** Hilgardia. 28(12):269-295. 1959.

Weis, S.A., et al. **A sensitive method for measuring changes in calcium concentration in 'McIntosh' apples demonstrated in determining effects of foliar calcium sprays.** J. Amer. Hort. Sci. 105:346-349. 1980.

In both the above works, the authors correlated disorders with low calcium levels in affected parts of the fruit. With careful, judicious sampling of certain sensitive fruiting parts it may be possible to obtain meaningful calcium analysis data.

page 55, lines 26-32
Stebbins, Bob. **Goodfruit growing in Oregon.** The Goodfruit Grower, Sept. 15, 1978. pp 42-43.

CHAPTER 10

page 57, lines 6,7

Bower, C.A. and L.M. Turk. **Calcium and magnesium deficiencies in alkali soils.** J. Amer. Soc. Agron. 38:723-727. 1946.

page 57, lines 24,25

Worley, R.E., R.L. Carter and A.W. Johnson, **Effect of magnesium sources and rates on correction of acute magnesium deficiency of pecan.** J. Amer. Soc. Hort. Sci. 100:487-490. 1975.

In this study dolomite lime applied to a soil with a pH of 5.7 was ineffective in correcting Mg deficiency while magnesium sulfate was effective.

CHAPTER 11

page 59, lines 6,7

Wallace, A. **Sulfuric acid may cure iron deficient crops.** Crops and Soils, March 1978, p 19-20.

With high lime soils it would be impractical to add enough sulfuric acid to completely neutralize the lime, however band application of sulfuric acid can make enough iron available to satisfy crop needs. Caution should be used with band application of acid since severe crop damage can occur if band is improperly placed. Certainly more work needs to be done in this area.

page 59, line 8

Gypsum aids sorghum quality. (A report of work done by Dr. Sterling Olsen), Agricultural Research (USDA publication) June 1978, p 15.

page 59, line 12

A relatively new test for peroxidase activity in plant tissue shows a good deal of promise as a diagnostic tool for iron deficiency. This test is described in HortScience, 13:284-285, **A rapid tissue test for diagnosing iron deficiencies in vegetable crops** by A. Bar-Akiva, D.N. Maynard and J.E. English. (1978).

CHAPTER 12

page 62, lines 9-12
Wallace, A. et al., **Effects of soil temperature and zinc application on yields and micronutrient content of four crop species grown together in a glasshouse.** Agron. J. 61:567-568. 1969.

page 63, bottom and page 64, line 1
Smith, M.W. et al. **Zinc and sulfur content in pecan leaflets as affected by application of sulfur and zinc to calcareous soils.** HortScience 15:77-78. 1980.

page 64, lines 2-4
Brown, A.L. et al. **Movement and availability of applied zinc.** Proceedings of Statewide Conference on Role of Micronutrients and Sulfur in Crop Production. University of California at Davis, Ca. Jan. 23, 24, 1964.

page 64, lines 27-34
Labanauskas, C.K. et al., **Low-residue micronutrient sprays for citrus.** Calif. Agric. Jan. 1973, p 6-7.

page 65, line 11
Foliar zinc sprays correct zinc deficiency in 1976. The Blue Anchor, March-April 1977, p 4.

page 66, lines 2-4
Smith, M.W. and J.B. Storey. **Zinc concentration of pecan leaflets and yield as influenced by zinc source and adjuvants.** J. Amer. Soc. Hort. Sci. 104:474-477. 1979.

CHAPTER 13

page 68, Boron deficiency
Boron analysis of fruiting parts is superior to analysis of leaves when diagnosing boron deficiency on some crops, esp. when abnormalaties occur on the fruiting parts; see Shear, C.B. and M. Faust. **Nutritional ranges in deciduous tree fruits and nuts.** Horticultural Reviews. 2:142-163. 1980.

page 68, lines 24-30
Chaplin, M.H. R.L. Stebbins and M.N. Westwood, **Effect of fall-applied boron sprays on fruit set and yield of 'Italian' prune.** HortScience 12:500-501. 1977.

Callan, N.W. et al. **Fruit set of 'Italian' prune following fall folar and spring foliar sprays.** J. Amer. Soc. Hort. Sci. 103:253-257. 1978.

Tukey, R.B. **Nutrient spray issue still controversial.** The Goodfruit Grower, April 1, 1971, p 25.

Burkhart, D.J. **Mid-Columbia pears may need fall boron sprays.** The Goodfruit Grower. October 15, 1978, p 6.

page 69, lines 16,17
Prather, R.J. **Sulfuric acid as an ammendment for reclaiming soils high in boron.** Soil Sci. Soc. Amer. Proc. 41:1098-1101. 1977.

page 69, lines 18-20
Ryan, J., S. Miyamoto and J.L. Stroehlein. **Relation of solute and sorbed boron to the boron hazard of irrigation water.** Plant and Soil, 47:253-256. 1977.

page 69, lines 21,22 & table
Reeve, E. and J.W. Shive. **Potassium-boron and calcium-boron relationships in plant nutrition.** Soil Science 57:1-14. 1944.

CHAPTER 14

page 70, lines 5-7
Rogers, E.G. Johnson and D. Johnson. **Iron-induced manganese deficiency in 'Sungold' peach and its effects on fruit composition and quality.** J. Amer. Soc. Hort. Sci. 99:242-244. 1974.

page 71, lines 6,7
Johnson, C.M., G.A. Pearson and P.R. Stout. **Molybdenum nutrition of crop plants. II. Plant and soil factors concerned with molybdenum deficiencies in crop plants.** Plant and Soil. 4:178-196. 1952.

page 71, lines 15-17
James D.W. **Soil fertility relationships of sugarbeets in central Washington: phosphorus, potassium-sodium and chlorine.** Technical Bulletin 69, Washington Agric. Exp. Sta., Pullman, Wash. July 1972.

CHAPTER 15

page 73, lines 7,8
Edwards, J.H., B.D. Horton and H.C. Kirkpatrick. **Aluminum toxicity symptoms in peach seedlings.** J. Amer. Soc. Hort. Sci. 101:139-142. 1976.

page 73, lines 20,21
Soane, B.D. and D.H. Saunder. **Nickel and chromium toxicity of serpentine soils in Southern Rhodesia.** Soil Sci. 88:322-330. 1959.

page 73, lines 26,27
Bradford, G.R. **Lithium in California's water resources.** California Agriculture, May 1963, p 6-8.

Aldrich, D.G., A.P. Vanselow and G.R. Bradford. **Lithium toxicity in citrus.** Calif. Agric. Oct. 1951, p 6.

page 74, lines 1,2
Soane, B.D. and Saunder (see above chromium reference).

Mishra, D. and M. Kar. **Nickel in plant growth and metabolism.** Botanical Review, 40:395-452. 1974.

page 74, lines 3,4
Cary, E.E. and W.H. Allaway. **Selenium content of field crops grown on selenite-treated soils.** Agron. J. 65:922-925. 1973.

page 74, lines 11,12
Wallace, A., E.M. Romney and R.T. Mueller. **Nitrogen-silicon interaction in plants growth in desert soil with nitrogen deficiency.** Agron. J. 68:529-530. 1976.

page 74, line 13
Okuda, A. and Takahashi. **The role of silicon,** p 123-146, in, The Mineral Nutrition of the Rice Plant, Proc. Symp. Intern. Rice Res. Inst., John Hopkins Press, Baltimore, U.S.A. 1965.

CHAPTER 16

page 75, table
Luckhardt, R.L. **Soil testing helps plan preplant nitrogen.** Agribusiness Fieldman, Sept. 1979, p 5. (Article covers U.C.'s Dr. Kent Tyler's work on soil nitrate).

A recent work indicates that soil nitrate data can be pushed through an equation to give a more refined nitrogen requirement index (NRI) for some crops. See, S. Roberts, et al. **Use of the nitrate soil test to predict sweet corn response to nitrogen fertilization.** Soil Sci. Soc. Amer. J. 44:306-308. 1980.

page 75, lines 20-23
For row crops, taking soil samples for nitrate analysis after each irrigation and correlating the results with petiole nitrate levels will give a firmer grip on N nutrition.

page 76, lines 30-32
Jones, J.B. Jr., **Should we, or shouldn't we standardize soil testing.** Crops & Soils. Feb. 1973, p 15-17. (A good discussion on the pros and cons of standardization).

page 76, last line and page 77, lines 1-4
A recent U.C. study shows that U.C. is changing this opinion; see, Uriu, et al. **Potassium fertilization of prune trees under drip irrigation.** J. Amer. Soc. Hort. Sci. 105:508-510. 1980.

page 77, quote
Sparks, Darrell. **Predicting the nutrient needs of pecan - a review.** Pecan South, Nov. 1978, p 280-283.

CHAPTER 17

page 79, bottom line and page 80, line 1
Luckhardt, R.L., **Plant analysis: where are we?** Agribusiness Fieldman, Feb. 1980, p 4.

In a recent review, two USDA scientists acknowledge the limitations of plant analysis as a tool for making precision fertilizer recommendations for tree crops: "Unfortunately, predicting the exact amount of nutrient required to bring the tree into nutritional balance is not simple. The method of trial and error must still be used." see Shear, C.B. and M. Faust. **Nutritional ranges in deciduous tree fruits and nuts.** Horticultural Reviews. 2:142-163. 1980. These authors also stress that fruit analysis can be superior to leaf analysis in diagnosing B and Ca deficiencies.

page 80, lines 3-8

Summer, M.E. **A new approach for predicting nutrient needs for increased crop yields.** Solutions. Sept.-Oct. 1978, p 68, 71, 74, 78.

Summer, M.E. **Interpretation of foliar analyses for diagnostic purposes.** Agron. J. 71:343-348. 1979.

Beaufilis, E.R. **Diagnosis and recommendation integrated system (DRIS).** Soil Science Bul. 1, University of Natal, Pietermaritzburg, South Africa. 1973.

page 81, line 1

Western Fertilizer Handbook (see General Reference section) has plant sampling guides.

page 81, lines 13-16

Terman, G.L. et al. **Nitrate-N and total-N concentration relationships in several plant species.** Agron. J. 68:556-560. 1976.

page 82, lines 13,14

Western Fertilizer Handbook (see General Reference section) has published interpretation guidelines. Other publications have similar guidelines.

page 82, lines 22-24

McClellan, G.W. **K for quality alfalfa.** Agrichemical West, Sept. 1967, p 6, 7, 8, 13.

McClellan, G.W. **Potassium's spectacular economic benefits.** Agrichemical West, Oct. 1968, p 6, 8, 10, 12, 14, 16.

Interestingly, McClellan used K/N, P/N and P/K ratios in his work, the same ratios used in the DRIS system.

page 83, lines 3-5

Childers, N.F. **Fruit Nutrition of Crops** (see General Reference section), p. 788-790 (Chapter on Grape nutrition by J.A. Cook).

page 83, lines 26,27

Sparks, D. **Nitrogen Scorch and the pecan.** Pecan South. Sept. 1976. 3:500-501. 1976.

page 83, lines 29,30

Chaplin, M.H., R.L. Stebbins and M.N. Westwood. **Effect of fall-applied boron sprays on fruit set and yield of 'Italian' prune.** HortScience. 12:500-501. 1977.

CHAPTER 18

page 85, bottom
> Luce, Bill. **Bill Luce Says.** Goodfruit Grower, May 1, 1977, p 4. (Luce, a highly respected orchard authority in the pacific northwest, discusses the benefits of foliar nutrient sprays applied from pink bud to petal fall).

page 86, lines 6-9
> Chaplin, M.H. et al. **Effect of fall-applied boron sprays on fruit set and yield of 'Italian' prune.** HortScience, 12:500-501. 1977.

page 86, lines 10,11
> Beeler Dick. **The joys and sorrows of foliar feeding.** Agrichemical Age. July-Aug. 1978. p 8, 17E.

page, 86, 88,
> Most of the information on cuticular and stomatal entry was taken, some of it verbatim, from 2 articles by Larry Eddings: **Pour it through the pores,** Calif. Farmer. 226:18, (Feb. 4, 1967), and **Foliar Nutrition,** a mimeo from Chevron Chemical Co., 575 Market St., San Francisco, Ca 94105. Edding's M.S. thesis (from the University of California at Davis, 1967) was on foliar nutrition and is a significant contribution to the body of knowledge on the subject.

page 89, lines 8,9
> Leonard, C.D. **Solvent boosts impact of foliar sprays.** Crops & Soils, Aug.-Sept. 1966, p 16.

> Leece, D.R. and J.F. Dirou. **Comparism of urea foliar sprays containing hydrocarbon or silicone surfactants with soil-applied nitrogen in maintaining the leaf nitrogen concentration of prune trees.** J. Amer. Soc. Hort. Sci. 104:644-648. 1979.

page 90, lines 8-10
> **Foliar fertilizers "green up" a crop, but don't raise yields.** Crops & Soils, April-May 1968, p 31.

page 90, line 13
> Proebsting, E.L. **Nitrogen sprays.** Calif. Agric. March 1951, p 12.

page 91, lines 13-15
> Shim K., J.S. Titus and W.E. Splittstoesser. **The fate of carbon and nitrogen from urea applied to foliage of senesc-

ing apple trees. J. Amer. Soc. Hort. Sci. 98:360-366. 1973.

CHAPTER 19

page 92, lines 24-28

Elliot, L.F. and F.J. Stevenson (Editors). **Soils for management of organic wastes and waste waters.** American Society of Agronomy publication (book). 1977.

The above reference represents the proceedings of a 1975 symposium. Sewage and waste disposal is a rapidly evolving subject; there is and will be alot of literature forthcoming on this subject.

page 94, lines 3-5

Guttay, A.J. **Fungus helps plants grow.** Crops & Soils, Aug.-Sept. 1975, p 14-15.

Menge, J.A., et al. **The effect of two mycorrhizal fungi upon growth and nutrition of avocado seedlings grown with six fertilizer treatments.** J. Amer. Soc. Hort. Sci. 105:400-404. 1980.

CHAPTER 20

page 96, lines 17-21

These ideas (skepticism toward the printed word) are taught by Dr. H.M. Reisenauer in his course on soil fertility at U.C., Davis - useful teaching.

page 97, lines 7-10

In most work, we search for little truths - occassionally a larger Truth will be discovered. That holiest of grails, THE TRUTH will always be, and should always remain beyond our grasp, although Einstein, perhaps, got a partial lock on TRuth; the pursuit, however, is more important than the conquest.

page 97, bottom (quote)

Grierson, W. **The enforced conservatism of young horticultural scientists.** HortScience. 15(3):228-229. 1980.

page 98, quote

Kondracke, M.H. **The trouble with lawyers.** The New Republic. 178(20):5-6. May 20, 1978.

page 99, quote
Day, Boysie E. **The morality of agronomy.** In: Agronomy in today's society. American Society of Agronomy, Special publication No. 33. 1978. pp 19-27.

GENERAL REFERENCES

1. **Advances in Agronomy.** Annual publication (in book from) prepared under the auspices of the American Society of Agronomy. Academic Press, Inc., 111 Fifth Ave., N.Y., N.Y. 10003. (This annual book is a review publication that covers approximately 10 subjects each year; often one or more nutritional subjects are included each year).

2. Buckman, H.O. and N.C. Brady. **The nature and properties of soils.** The MacMillan Co., N.Y. 1961+. (An introductory soils text containing a wealth of basic, sound information).

3. Chapman, H.D. (Ed.). **Diagnostic criteria for plants and soils.** University of California, 1966. (An excellent reference book with chapters on all the nutrients and including plant analysis levels for numerous crops).

4. Childers, N.F. (Ed.) **Fruit Nutrition (or Nutrition of Fruit Crops).** Rutgers University, New Brunswick, N.J. 1966. (Discusses nutrition by crop; includes plant analysis tables for various fruit crops).

5. Epstein, E. **Mineral nutrition of plants: principles and perspectives.** John Wiley & Sons, Inc., N.Y. 1972. (A more advanced text on plant nutrition; heavy on plant physiology).

6. Lorenz, O.A. and D.M. Maynard, **Knott's handbook for vegetable growers,** 2nd edition. John Wiley & Sons, Inc., N.Y. (Contains nutrient level tables for vegetable crops plus a wealth of other useful information). 1980.

7. Mortvedt, J.J. (ed.). **Micronutrients in agriculture.** Soil Science Society of America, 677 S. Segoe Rd., Madison, Wis. 53711. 1972. (A 666 page book with individual chapters by specialists).

8. Reisenauer, H.M. (ed.). **Soil and plant-tissue testing in California.** Bull. 1879. Univ. of Calif. Agric. Ext. Service. 1976. (A 54 page manual. Although developed for California, much of it is applicable to other areas).

9. Olson, R.A. et al. (eds.). **Fertilizer technology & use.** Soil Sci. Soc. of Amer., 677 S. Segoe Rd., Madison, Wis. 53711. 1971 (2nd printing, 1973). (A 611 page book detailing most aspects of fertilizer use).

10. Soil Improvement Committee, Calif. Fertilizer Assn. **Western fertilizer handbook.** 5th edition, 1975. (A compact 250 page book containing a wealth of information; impressive, because it is a joint effort of the University of California and the California fertilizer industry).

11. **Solutions (also called Fertilizer Solutions).** A bi-monthly publication of the National Fertilizer Solutions Association, 8823 N. Industrial Rd., Perioa, Il. 61615. (At regular intervals this publication has run a series on the major and minor nutrients authored by University and industry personnel; this series provides a means of keeping up with changes in thinking about the various nutrients).

12. Tisdale, S.L. and W.L. Nelson. **Soil fertility and fertilizers.** The Macmillan Co., N.Y. 1966. (A comprehensive 694 page soil fertility text).

Note: A revised edition of USDA Handbook 60 **(Diagnosis and improvement of saline and alkali soils)** should be available around 1982 and should be added to the above list when available.